JESUS LIGHT *and* DESIGN

What Light Reveals about Creation, Information, and Reality

A Real Science Radio *Teaching Series Book*

Jesus, Light & Design

Published by
Real Science Radio Press

Learn more at:
rsr.org
YouTube.com/@RealScienceRadio
store.rsr.org

979-8-9955602-0-3 paperback
979-8-9955602-1-0 hardcover
979-8-9955602-2-7 ebook

First Edition
Printed in the United States of America

This book is dedicated to **Bob Enyart**,
founder of Real Science Radio.

Bob has gone on to heaven now. He was the wisest person I have ever known, a fearless teacher, a brilliant thinker, and a tireless defender of truth.

His passion for Scripture, his love for science, his defense of the unborn, and his unwavering commitment to proclaiming the Gospel shaped countless lives, including my own.

Bob challenged assumptions, pursued truth wherever it led, and inspired others to see God's design in every aspect of creation.

Bob's legacy continues through the work of Real Science Radio and through all who were encouraged by his example.

His voice is deeply missed, but his influence endures.

"And the light shines in the darkness, and the darkness did not comprehend it."

JOHN 1:5

CONTENTS

“

I am the
light of
the world.

John 8:12

INTRODUCTION

Why Light Matters

From Genesis to Revelation, light takes center stage. When babies are born, they experience light.

God creates by light, Christ reveals Himself as light. Salvation is described as coming out of darkness into light. Light is also the final reality of the New Creation, where Scripture promises there will be no night because God Himself will be the light (Revelation 21:23-25).

Light is the first created reality that God explicitly declared "good" in Genesis 1, on the very first day of creation.

God brought light into our universe even before sources we normally associate with light such as the sun and stars. These were created on Day 4. Light therefore does not depend on the heavens, but rather the heavens serve the light God had already spoken into existence.

Modern science has only deepened this mystery. Physicists now recognize that light behaves in ways that challenge our assumptions about matter, causality, and even information. Light can act:

a. **like a wave,**
b. **like a particle, and**
c. **like a carrier of meaning.**

We move through light, measure with light, communicate with light, and depend on light for life. Yet light is far more than illumination. Light is a window into the deepest structure of reality. Darkness itself doesn't exist and is instead the absence of light. When you turn on a light, darkness flees.

Light is far more than illumination. Light is a window into the deepest structure of reality.

Astronomers attempt to read the history of the universe through light. Biologists find light-driven processes at the foundation of life. Engineers use light to transmit information across the globe and out into space. In every domain, light is central.

This book presents one guiding idea: Light is not accidental. It is designed, and it continually points beyond itself to its Designer. From the structure of the cosmos to the spark of life in the womb, from quantum experiments to the glow of bioluminescent seas, light bears witness to:

- order,
- intelligence, and
- purpose.

The heavens truly declare the glory of God — and they do so in the language of light.

"

For with You is the fountain of life; In Your light we see light.

PSALM 36:9

CHAPTER 1

What Is Light?

Light is part of the electromagnetic spectrum. This vast family of energy includes:

- radio waves
- microwaves
- infrared light
- visible light
- ultraviolet
- X-rays
- gamma rays

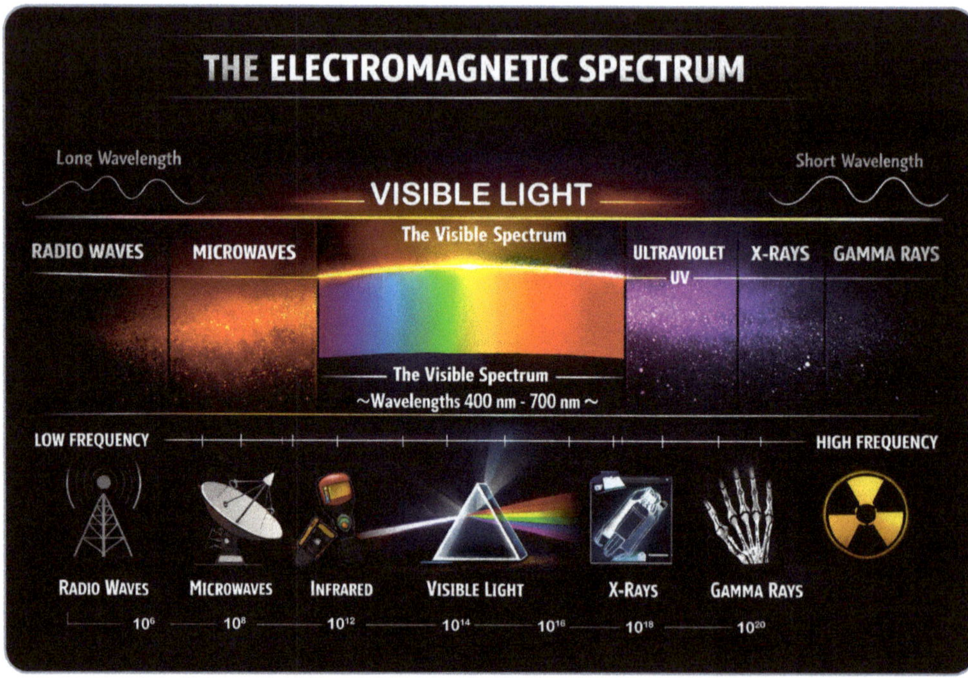

Visible light is only a tiny slice of this spectrum, yet visible light is precisely tuned for human vision.

The light visible to our eyes occupies only a tiny portion of a much larger range of electromagnetic radiation that includes radio waves, microwaves, infrared, ultraviolet, X-rays, and gamma rays.

A simple experiment shows that our eyes see only part of the light around us. If you point a television remote control at your phone's camera and press a button, you will often see a bright purple flash on the phone screen, even though you see nothing with your eyes. The remote emits **infrared light**, which is invisible to humans but detectable by most smartphone cameras.

In a similar way, bees can see **ultraviolet light**, revealing patterns in flowers, often called **nectar guides,** that are completely hidden from our human eyes. These patterns direct bees to the flower's nectar. This bee sight reminds us that reality is far richer than what our senses directly perceive.

The light we see is real, but it is only a small window into a much larger, invisible world of electromagnetic energy. If our eyes saw a different part of the spectrum, the world would look utterly alien. But the light God created is perfectly matched to human perception, just as Earth is perfectly suited for life.
We know that Light

- carries energy.
- warms the Earth,
- powers photosynthesis,
- drives the water cycle, and
- sustains ecosystems.

> Reality is far richer than what our senses directly perceive.

Remove sunlight from Earth and within weeks most ecosystems would collapse, showing how intimately life depends on light.

In laboratories, devices like radiometers demonstrate that light can produce motion.

Modern solar panels convert sunlight directly into electricity to power homes, satellites, and spacecraft.

Engineers have also built solar sails, large reflective sheets that can be pushed through space by sunlight alone, proving that light can literally propel objects across the cosmos. Light truly does real work in the physical world.

These examples remind us that light is not merely something we see. It is an active force capable of moving matter and powering technology.

Light doing work: Photons striking the black vanes cause the radiometer to spin.

Sunlight powering a space station: Solar arrays on the International Space Station convert sunlight into electricity.

Solar Sails: IKAROS is the first spacecraft to successfully demonstrate solar sail technology in interplanetary space.

Yet light is also deeply mysterious:

Light has no mass at rest.

Light is theorized to always travel at the same speed in a vacuum.

Nothing is known to outrun light.

Nothing is understood to slow light down in empty space.

Where there is light, darkness flees.

Even with its mysteries, light is reliable. It does not waver or fluctuate in empty space.

When you think of the stability of light, the "Light of the world" comes to mind, Jesus Christ (John 8:12). In a sense, light reflects the dependable character of God.

"Every good gift and every perfect gift is from above, and comes down from the Father of lights, with whom there is no variation or shadow of turning." — ***James 1:17***

Light also reveals, it makes things visible. Without light, there is no seeing, no reading, no color, no form, no clarity. In Scripture, light represents everlasting power, boundless beauty, infinite wisdom, and unchanging truth. In both science and Scripture, light is foundational.

NOTICE AND DISCUSS

1. ***What feature of light surprised you or changed the way you think about it?***
2. ***What function of light helps you do your daily stuff?***
3. ***Why do you want to be a person of light, rather than darkness?***

“

God asked Job
a remarkable
question:
“By what
way is light
diffused?”

JOB 38:24

CHAPTER 2

The Beauty of Light

DESIGN, COLOR, AND PERCEPTION

This question points to one of the most beautiful and mysterious features of creation. Light is not merely useful to illuminate; light also creates beauty.

Light fills the world with color, brilliance, and visual wonder. The beauty we experience in sunsets, rainbows, flowers, oceans, and living creatures all depends on the remarkable properties of light.

WHITE LIGHT AND THE HIDDEN SPECTRUM

In the 1600s, discoverer of optics, mathematician, and scientist Isaac Newton performed a famous experiment that revealed a profound truth about light. Using a triangular block of glass called a prism, he allowed a beam of sunlight to shine through his window and pass through the glass. On the opposite side of the prism, a brilliant band of colors appeared.

Newton uses a prism to show the multicolored composition of white light

Newton demonstrated that ordinary white light is not simple. White light is composed of multiple colors — red, orange, yellow, green, blue, indigo, and violet — each corresponding to a different wavelength of light.

This band of colors is called the visible spectrum. It is part of the much larger **electromagnetic spectrum**, which includes all forms of electromagnetic radiation such as radio waves, infrared, ultraviolet, X-rays, and gamma rays. This spectrum was introduced in Chapter 1 and now we go a bit deeper into it.

A prism separates white light into its component wavelengths because each wavelength bends at a slightly different angle as it passes through the glass. What appears to our eyes as pure white light is actually a rich combination of distinct frequencies producing hidden hues that remain unseen until revealed.

Beyond the visible spectrum lie infrared and ultraviolet radiation, invisible to us yet equally real. Human vision is therefore tuned to a specific window of reality, precisely matched to the light that surrounds us.

This raises an important insight: the part of the world we see is shaped by the properties of light and God's design of our senses.

The band of colors revealed by a prism is called the spectrum. These are colors the human eye can see. Memorize them in order with Roy G. Biv:

R.ed
O.range
Y.ellow
G.reen
B.lue
I.ndigo (purply blue)
V.iolet

WHAT IS COLOR?

Surprisingly, light itself is not colored. Color is not a physical substance and is not an inherent property of light itself. Color is determined by wavelength and interpreted by the human visual system. Each wavelength of light stimulates the eye and brain in a specific way, producing the sensation we call color. What we perceive as red, blue, or green is a response of the mind to particular frequencies of electromagnetic radiation. Color therefore exists at the intersection of physical reality and human perception.

The universe produces ordered wavelengths of light, and the human eye is structured to interpret them as meaningful visual experiences. Light and perception work together in a finely tuned system that allows us to experience the beauty of creation.

THE DESIGN OF VISION – LIGHT AND THE HUMAN EYE

The remarkable properties of light would mean little if living organisms were not equipped to detect and interpret it. Yet the biological systems that perceive light display extraordinary precision. The human visual system is not a simple detector but a highly sophisticated optical and information-processing system that converts photons into meaningful perception.

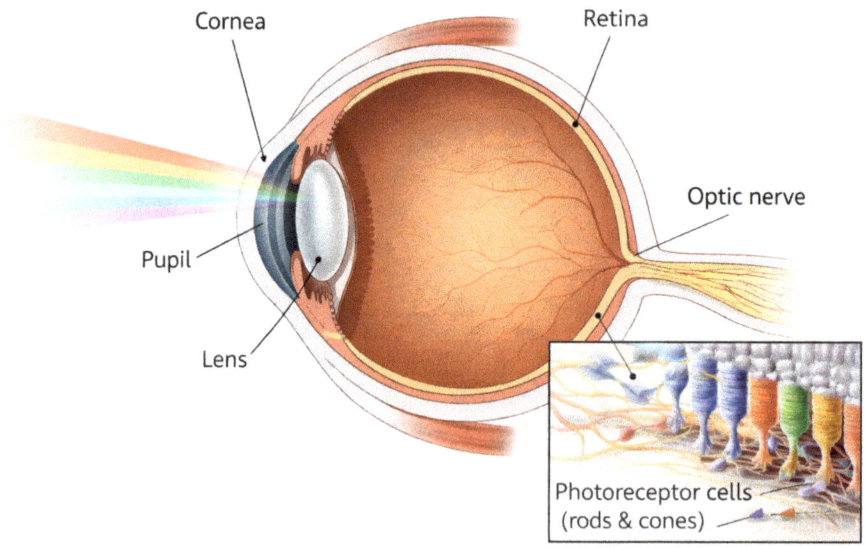

The eye is carefully structured to receive, focus, and interpret light with astonishing efficiency.

Light entering the eye passes through the cornea and lens and is focused onto the retina, where specialized photoreceptor cells (rods and cones) convert photons into electrical signals that travel through the optic nerve to the brain.

DETECTION NEAR THE LIMITS OF PHYSICS

The retina, located at the back of the eye, contains specialized cells called photoreceptors that respond to incoming light. Some of these cells, known as rods, are so sensitive they can detect individual photons under extremely low-light conditions. Experiments have shown that the human visual system operates near the theoretical limits allowed by quantum physics. Few engineered detectors approach this level of sensitivity.

This ability allows vision under starlight and extremely dim illumination. The eye is not merely functional. The eye-optic nerve-brain combination operates with extraordinary efficiency at the physical limits of what is possible.

OPTICAL PRECISION AND CORRECTING TO MAINTAIN FOCUS

The eye in the human vision system contains an intricate optical system. The transparent cornea and lens bend incoming light to form a focused image on

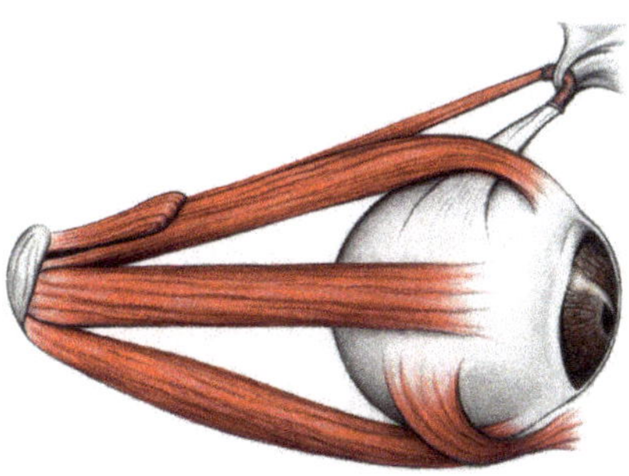

EYE PULLEY SYSTEM – GOD'S REMARKABLE DESIGN

One of the muscles that moves the eye does something remarkable. The **superior oblique muscle** passes through a tiny cartilage loop called the *trochlea*, functioning like a precision **pulley** that redirects force to control eye rotation.

Even more remarkable, this complex pulley system is **duplicated and mirrored on both sides of the head**, providing symmetrical control of both eyes.

How could random mutations gradually construct a working pulley involving multiple tissues – muscle, cartilage, nerve, connective structures – and then duplicate and coordinate it on both sides? They couldn't! The eye testifies of God!

the retina. The curvature of the lens continuously adjusts to maintain focus on objects at different distances. Muscles surrounding the eye automatically track motion, stabilize images, and maintain alignment.

Even more remarkable, the optical design of the eye-brain combo includes mechanisms that compensate for distortions and imperfections. The brain continually corrects blind spots, stabilizes motion, and interprets incomplete visual information into coherent images.

What we experience as effortless sight depends on constant, complex processing.

VISION TUNED TO THE VISIBLE SPECTRUM

Human vision is based on three types of cone cells, each sensitive to different ranges of wavelength (color) within the visible spectrum. These receptors respond primarily to red, green, and blue light, allowing the brain to construct the full range of perceived colors.

This three-channel system is precisely matched to the spectral range of sunlight that reaches Earth's surface, another testimony to God's amazing creativity! The wavelengths most abundant in our environment correspond to the wavelengths our eyes detect most efficiently.

Light, the environment, and biological perception are closely aligned.

Rather than a simple camera, the eye performs extensive information processing.

TRANSPARENCY AND ORDER

For vision to function, multiple structures within the eye must remain nearly perfectly transparent. The cornea, lens, and internal fluids allow light to pass with minimal scattering. Even slight irregularities in this transparency significantly degrade vision. The maintenance of this optical clarity requires complex biological regulation. For example, cataracts occur when the normally transparent lens becomes cloudy, scattering light and preventing sharp images from forming on the retina. This illustrates how delicate the eye's optical system is and how precisely its transparency must be maintained.

The organization of retinal cells also exhibits remarkable order. Millions of photoreceptors, specialized cells that detect light, convert light into electrical signals, which are processed by layered neural circuits before transmission to the brain. Rather than a simple camera, the eye performs extensive information processing before the signal ever leaves the retina to travel to the brain.

LIGHT INTERPRETED AS MEANING

The brain does not merely receive light, it interprets it. Electrical signals from the retina are processed by the brain into depth, motion, color, and form. The visual system reconstructs three-dimensional space, identifies patterns, and recognizes objects in fractions of a second.

Light becomes perception.

Energy becomes meaning.

The physical properties of light, the structure of the eye, and the processing power of the brain operate together as an integrated system. The ability to perceive the world through light reflects not only biological function but coordinated design.

If light reveals order in the universe through sight, the system that perceives light reflects that same order within living organisms.

LIGHT EXPERIENCED AS BEAUTY

The beauty of light is not limited to simple color. Many of the most breathtaking displays in nature arise from complex interactions between light and matter.

Rainbows form when sunlight refracts and reflects through water droplets in the atmosphere.

Sunsets blaze with color as Earth's atmosphere scatters shorter wavelengths and allows longer wavelengths to dominate the sky.

Crystals, gemstones, and oceans all display unique optical effects because of how they interact with light.

These phenomena reveal that God has structured the physical world in ways that produce beauty.

Light is not random radiation. It behaves according to precise laws that generate order, harmony, and aesthetic experience.

LIGHT IN THE MORPHO BUTTERFLY

One of the most stunning examples of light's beauty appears in butterflies. The brilliant colors of certain butterflies, especially the vivid blues seen in tropical species such as Morpho butterflies, are produced by the structure of their wings, not by pigments.

Butterfly wings consist of a thin, translucent membrane covered with thousands of microscopic scales. These scales are flattened outgrowths of individual cells. Typically about 100 micrometers long (about one-tenth of a millimeter), these scales overlap like tiny roof tiles. Under magnification, they reveal intricate patterns of grooves and ridges. They form layered ridges that resemble tiny tree-like structures when viewed side by side.

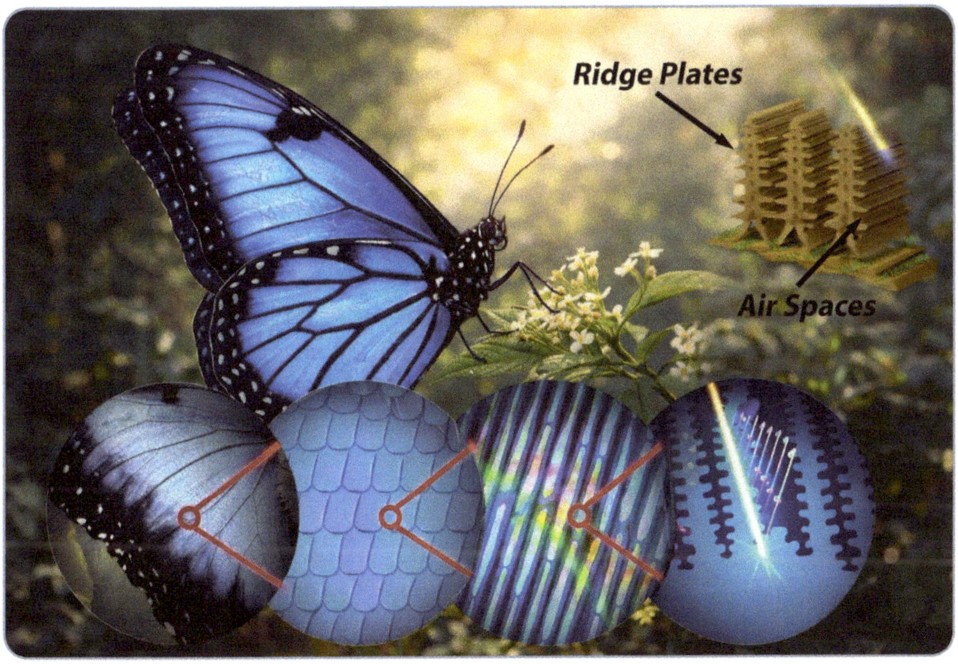

The intricacy of these structures is remarkable. Many of the features responsible for butterfly colors occur at the nanoscale, so small that they approach the resolution limits of conventional optical microscopes.

What is amazing is we see the vivid blues of Morpho butterflies because these structures on their wings create the colors through a phenomenon known as *interference*—light through different layers. When waves of light reflect from different surfaces of the wing's microscopic structures, they can combine constructively or destructively. If the peaks and troughs of the waves align, they reinforce one another, producing intense color. If they are out of alignment, they cancel out.

In some butterflies, the spacing of these microscopic structures selectively reinforces specific wavelengths of light. For example, structures spaced about 200 nanometers apart strongly reflect blue wavelengths (roughly 400–480 nanometers) while other wavelengths are absorbed by underlying pigments. The result is a vivid, shimmering blue that can flash brilliantly in sunlight. Nanometers are billionths of a meter.

Pilots flying over tropical rainforests have reported seeing brief flashes of blue as large groups of butterflies move above the forest canopy, a remarkable display produced entirely by the interaction of light and microscopic structures.

This interference often produces a visual effect known as iridescence, where colors appear to change when viewed from different angles. You might recognize the word *iridescent*, which describes surfaces that shimmer with shifting colors. Iridescence demonstrates how living systems harness the wave nature of light to produce extraordinary beauty.

THE FINE-TUNING OF LIGHT

The beauty produced by light depends on remarkable precision:

- Exact wavelengths of radiation +
- Precise optical properties of materials +
- Highly ordered microscopic structures +
- Visual systems capable of interpreting color.

The universe is structured so light interacts with matter in ways that produce harmony and visual richness.

Human perception is structured to recognize and appreciate that beauty. Light reveals not only objects but order. It reveals pattern, symmetry, and design.

Another remarkable feature of light is that its properties appear precisely suited for life. Modern physics has revealed that the behavior of light depends on fundamental constants of nature, fixed values that determine how strongly forces act, how atoms form, and how energy interacts with matter. If these values were even slightly different, the universe as we know it would not exist.

Among the most important of these constants is the strength of the electromagnetic force, which governs the behavior of light and its interaction with matter. Physicists describe this strength using a number called the **fine-structure constant**, commonly represented by the Greek letter alpha (α). Its value is approximately:

$$\alpha \approx 1/137$$

This number determines the strength of electromagnetic interaction between charged particles. It governs how tightly electrons bind to atoms, how atoms form molecules, how light is emitted and absorbed, and how chemistry itself operates.

If this value were only slightly different, stable matter would not exist. A stronger electromagnetic force would bind electrons too tightly to atomic nuclei, preventing the chemical reactions necessary for life.

A weaker force would prevent atoms from holding together at all. The structures required for stars, planets, chemistry, and biology all depend on this precise balance.

This fine-tuning extends beyond abstract constants to the specific wavelengths of light that surround us. As mentioned in Chapter 1, human vision is limited to a narrow band of the electromagnetic spectrum known as visible light. At first glance this might seem accidental, but deeper study reveals a remarkable convergence of conditions that point to creative design, and God as its Creator.

Our Sun emits most of its energy in the visible portion of the spectrum. Earth's atmosphere is highly transparent to these same wavelengths, allowing sunlight to reach the surface with minimal obstruction.

Water, which makes up most of Earth's surface and most living cells, is also unusually transparent to visible light, allowing light to penetrate oceans and sustain aquatic ecosystems through photosynthesis. Photosynthesis is the process by which plants and algae use light energy to produce food. These three facts align in a striking way:

- The Sun produces its strongest radiation in visible wavelengths
- Earth's atmosphere allows those wavelengths to pass through efficiently
- Water transmits those same wavelengths to sustain life

The window of visible light is therefore uniquely suited to life on Earth. It provides energy for biological processes, enables vision, and allows the environment to remain stable. Outside this narrow band, radiation becomes either too energetic and destructive or too weak to drive essential chemical reactions.

The universe therefore appears calibrated at a fundamental level—strong evidence that a benevolent and personal creator was behind it all.

The laws governing light and electromagnetism are not arbitrary; they exist within an extremely narrow range that permits order, structure, and life.

GOD'S QUESTION ABOUT DIFFUSED LIGHT

God's question to Job, "By what way is light diffused?" reminds us that even today, light remains a profound mystery. We can imperfectly describe how light behaves, measure its properties, and harness its power, yet its deeper nature continues to challenge scientific understanding.

Even Albert Einstein admitted, "All the fifty years of conscious brooding have brought me no closer to the answer to the question, 'What are light quanta?'"

Because the behavior of light lies at the heart of quantum theory, physicist Richard Feynman likewise wryly observed, "I think I can safely say that nobody understands quantum mechanics."

Light spreads, reflects, refracts, interferes, and reveals. It carries energy and information while simultaneously producing beauty.

The world is filled with color because of the properties of light. Living organisms display intricate optical structures that interact with light in astonishing ways. We see because of light, our eyes, and our brain.

Human perception is tuned to interpret these interactions as meaningful experiences.

The beauty of light is not accidental. God built light into the structure of reality.

LIGHT AS A WINDOW INTO CREATION

From the spectrum revealed by a prism to the shimmering wings of a butterfly, light displays order, precision, and elegance. Light transforms invisible radiation into visible beauty and allows human beings to experience the richness of creation.

Light reveals that the universe is not merely functional but expressive. The universe is filled with splendor, structure, and meaning. The beauty of light invites us to wonder not only at what we see, but at the source of that beauty. The universe prepares us to venture into the deeper mysteries of light.

In our next chapter we turn to one of the most surprising discoveries in science: the threefold nature of light itself.

NOTICE AND DISCUSS

1. *What do you like about the hidden hues in white light? What might this reveal about God's creativity?*
2. *What amazes you most about how the human eye, brain, and light work together to produce vision?*
3. *Why is light a window into God's creation?*

“

For there are three that bear witness in heaven: the Father, the Word, and the Holy Spirit; and these three are one.

1 John 5:7

CHAPTER 3

The Triality of Light

WAVE, PARTICLE, INFORMATION

For centuries, scientists argued about a simple question: What is light?

In the 1600s, Isaac Newton proposed that light consists of tiny particles traveling from place to place.

Around the same time, mathematician and physicist Christiaan Huygens argued that light behaves as a wave spreading through space like ripples on water.

In the early 1800s, new experiments seemed to settle the debate. Thomas Young, a scientist knowledgeable on many subjects, demonstrated that light produces interference patterns, a behavior expected of waves.

Later in the century, physicist and mathematician James Clerk Maxwell showed that light is an electromagnetic wave, unifying electricity, magnetism, and optics into a single theory. By the late 1800s, most scientists believed the answer was clear: light is a wave.

However, more recent discoveries revealed that light also behaves like a particle. In 1905, Albert Einstein showed that light transfers energy in discrete packets, later called photons. Experiments supported one description and then the other. Each explained important features of light, yet neither told the whole story.

Modern physics arrived at a startling conclusion:

Light is both wave and particle.
Depending on how we observe it.
Or measure it.

Wave and Particle

At first, light appears to behave like a wave. It spreads out, overlaps with itself, and produces patterns of bright and dark bands. But when scientists examined light more closely, they discovered something far more mysterious.

In the famous double-slit experiment, light passes through two narrow slits and produces an interference pattern, exactly what waves do. Yet when a detector is placed at one or both slits to measure which path the light takes, the interference pattern disappears, and the result becomes a particle-like distribution.

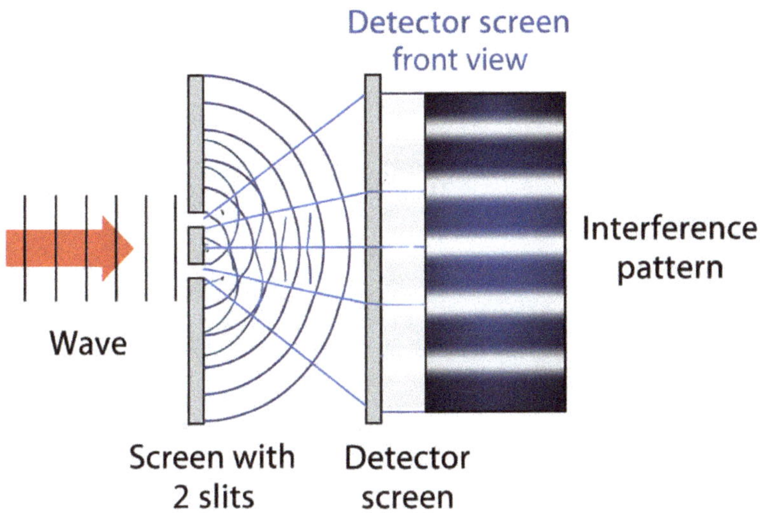

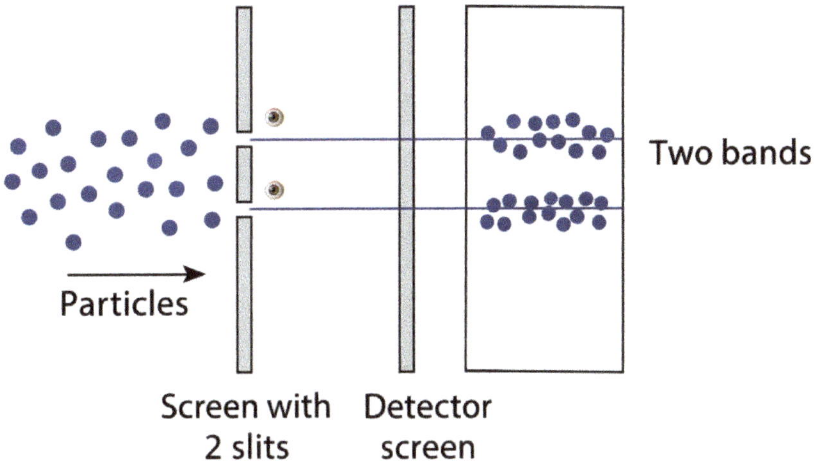

With an observer at the two slits, discrete lines appear.

When scientists detect individual photons, light arrives as localized packets of energy. Each photon strikes a specific point on the detector. Even in the double-slit experiment, the wave pattern builds up one particle-like detection at a time.

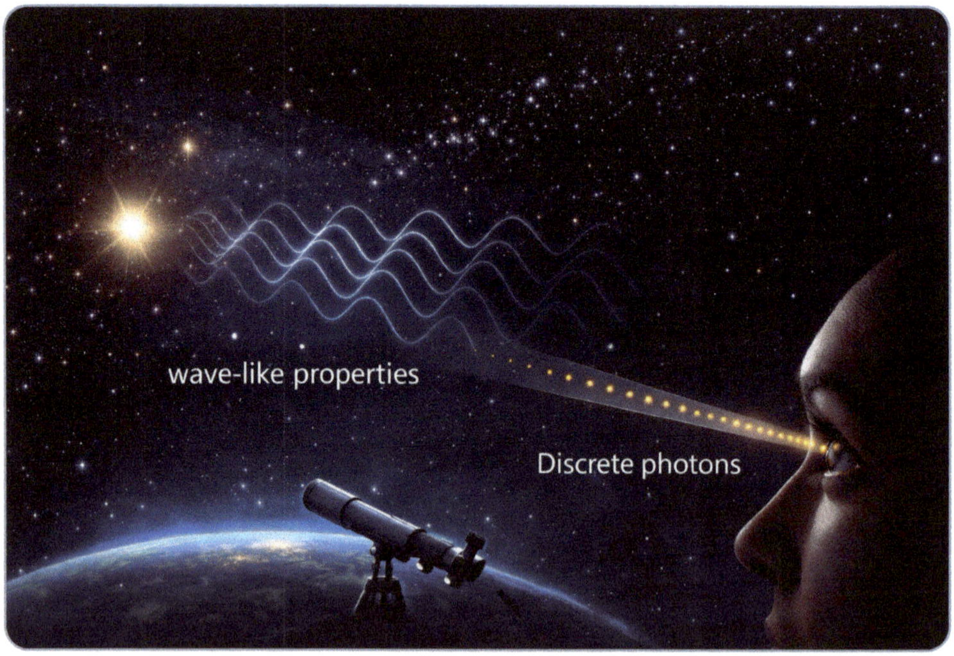

This dual behavior of light is not limited to laboratory experiments. Light from distant stars provides another example. As it travels through space it exhibits wave-like properties, but when it is detected by a telescope or by our eyes, light is absorbed in discrete packets of energy called photons.

This raises a profound question: Did God design light like this so we can see distant stars? The mystery of the quantum world continues to boggle the mind. Recall that physicist Richard Feynman once remarked, "I think I can safely say that nobody understands quantum mechanics."

Light refuses to fit into a single category. It behaves as a wave under some conditions and as a particle under others. Physicists call this *wave–particle duality*, one of the central discoveries of quantum mechanics.

But this is not the end of the story.

A THIRD ASPECT: INFORMATION

Another feature of light is often overlooked. Light does not merely transport energy.

Light carries information.

When light enters your eyes, it conveys images.

When light travels through fiber-optic cables, it carries digital data. When astronomers study starlight, they decode information about temperature, composition, motion, and distance.

A beam of light is not just energy in motion. A beam of light is a structured signal. Its frequency, intensity, polarization, and pattern encode meaningful content.

The universe communicates through light. Even the most distant galaxies send messages through their spectra. Their light carries detailed information across vast distances of space. By spreading that light into its component colors, astronomers can read patterns that reveal what elements those galaxies contain and how they are moving through space.

Each chemical element produces a unique set of spectral lines, a distinctive pattern of wavelengths that serves as its signature. Hydrogen, helium, oxygen, and every other element leave their own recognizable marks in the light they emit or absorb. By studying these patterns, scientists can determine temperature, chemical composition, and information about the motion of stars relative to Earth.

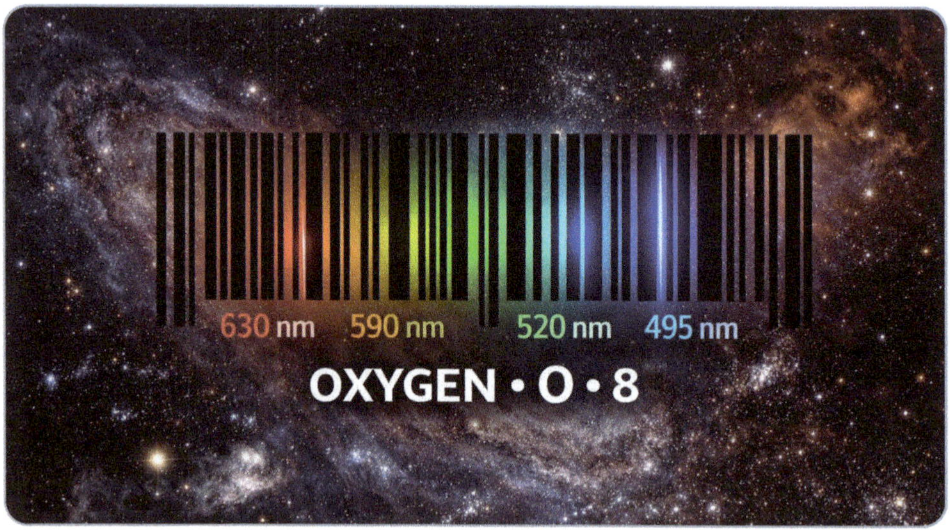

In this way, light functions as a cosmic messenger, faithfully carrying information across vast distances in the universe. Scientists often describe these spectral patterns like barcodes revealing what distant objects contain. The universe, in a very real sense, communicates through light.

This suggests something deeper:

Light has three aspects — wave, particle, and information.

Rather than a duality, in light we witness a **triality**.

EXACTNESS: A CLUE TO INFORMATION

Another remarkable feature of the physical world strengthens this discovery: numbers are exact. The number three never varies. It does not wear out or drift. It remains perfectly itself regardless of how often it appears.

Strangely, the same kind of exactness appears in fundamental physics.

Every electron is identical to every other electron.

Every proton is identical to every other proton.

Particles created in widely separated regions of the universe possess precisely the same properties.

The proton is about 1,835 times heavier than the electron, yet every proton appears exactly the same as all others. Its electrical charge is always equal in magnitude and opposite to the electron's charge. Experiments have repeatedly confirmed that particles of the same type are indistinguishable. Consider that human manufacturing always involves tolerances and variation. Yet nature appears to produce fundamental particles with perfect precision.

Why should that be?

The exactness of numbers is easy to understand as they are not physical objects. Instead, they are massless, abstract realities. But why should physical particles display the same kind of precision?

This has led some prominent physicists to suggest that the physical world reflects an underlying informational or mathematical structure. Over the history of modern physics, explanations have moved from particles, to fields, and increasingly toward information as fundamental. Physicist John Wheeler captured this shift in the phrase "it *from* bit," proposing that physical reality ultimately arises from information.

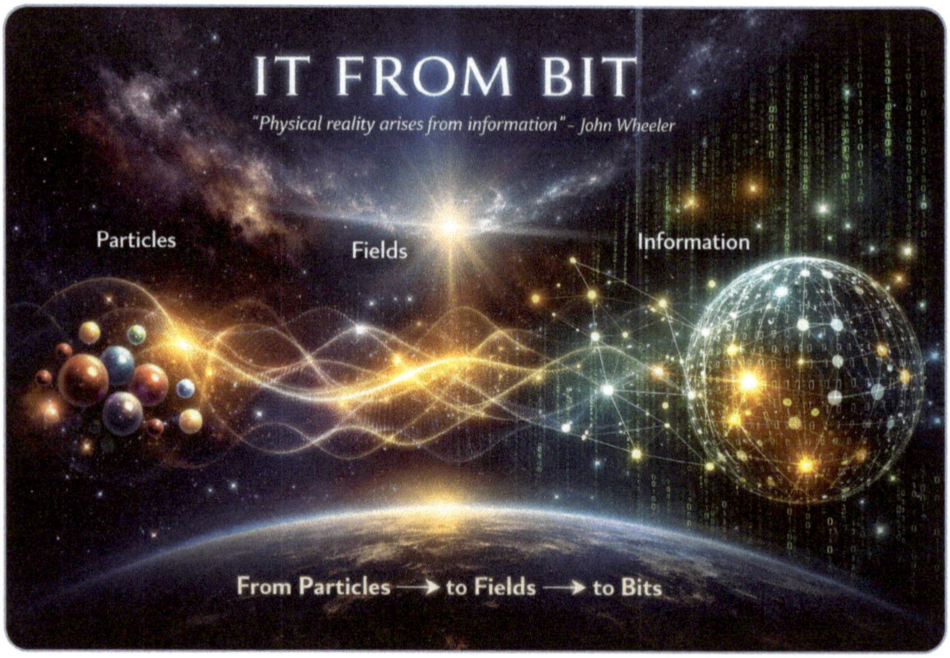

A statement is not made of ink or sound waves. Meaning rides on physical carriers but is not identical to them. In the same way, physical processes may carry deeper informational content.

Mathematician and cybernetics pioneer Norbert Wiener expressed a similar insight:

"Information is information, not matter or energy."

In other words, the universe is not simply matter and energy. It also reflects an underlying informational structure that governs how matter and energy behave.

The regularity and exactness of matter therefore point beyond mere mechanism to something deeper. A statement, after all, is not made of ink or sound waves. Meaning rides on physical carriers but is not identical to them. In the same way, physical processes may carry deeper informational content.

THE IDENTITY OF PARTICLES

Quantum physics reveals another striking feature of reality: particles of the same kind appear perfectly identical. Electrons, neutrinos, and quarks, some of the fundamental particles that make up matter, all share this property. Experiments demonstrate that even newly created particles match the properties of those formed centuries ago. The universe shows no evidence of gradual drift or variation at this fundamental level. This uniformity is difficult to explain purely in mechanical terms. Why should nature "manufacture" particles without variation? Why no tolerances? Why no slight deviations?

One possibility is that particles are not simply tiny physical objects but stable expressions of deeper laws, like mathematical values consistently expressed in different contexts. Just as the number three is always three, a fundamental particle may be the repeated expression of an underlying informational pattern.

This perspective does not deny physical reality. Rather, it suggests that physical objects are grounded in deeper structure, in rules, relationships, and information.

INFORMATION AND THE QUANTUM WORLD

Quantum mechanics increasingly points toward an informational view of reality.

Some physicists describe the quantum state as a representation of information rather than a physical wave. A particle's behavior is governed by probabilities, mathematical descriptions of possible outcomes. These probabilities are not physical objects; they are conceptual descriptions.

The famous *wave function*, described by physicist **Erwin Schrödinger's equation**, behaves like a mathematical entity. It encodes possibilities and relationships rather than tangible substance.

This helps explain puzzling quantum behavior. A photon can exhibit wave-like behavior across space and yet produce a single localized detection. The wave describes possibilities.

The particle expresses a specific outcome.

Information provides a natural bridge between wave and particle behavior.

FROM DUALITY TO TRIALITY

If light exhibits wave behavior, particle behavior, and informational structure, then perhaps we are not dealing with a duality at all.

Perhaps light is best understood as a unity expressed in three complementary ways:

- **Wave** — spreading patterns and probabilities;
- **Particle** — localized energy transfer;
- **Information** — structured meaning and order.

These are not competing explanations but different aspects of one reality.

Physicist Richard Feynman observed that quantum mechanics can be described using multiple distinct mathematical frameworks, each valid yet conceptually different.

Reality seems richer than any single model.

The universe may be fundamentally relational and informational, not merely mechanical.

A TRIUNE PATTERN IN CREATION

This threefold structure is striking. Throughout creation we see patterns of unity expressed in plurality.

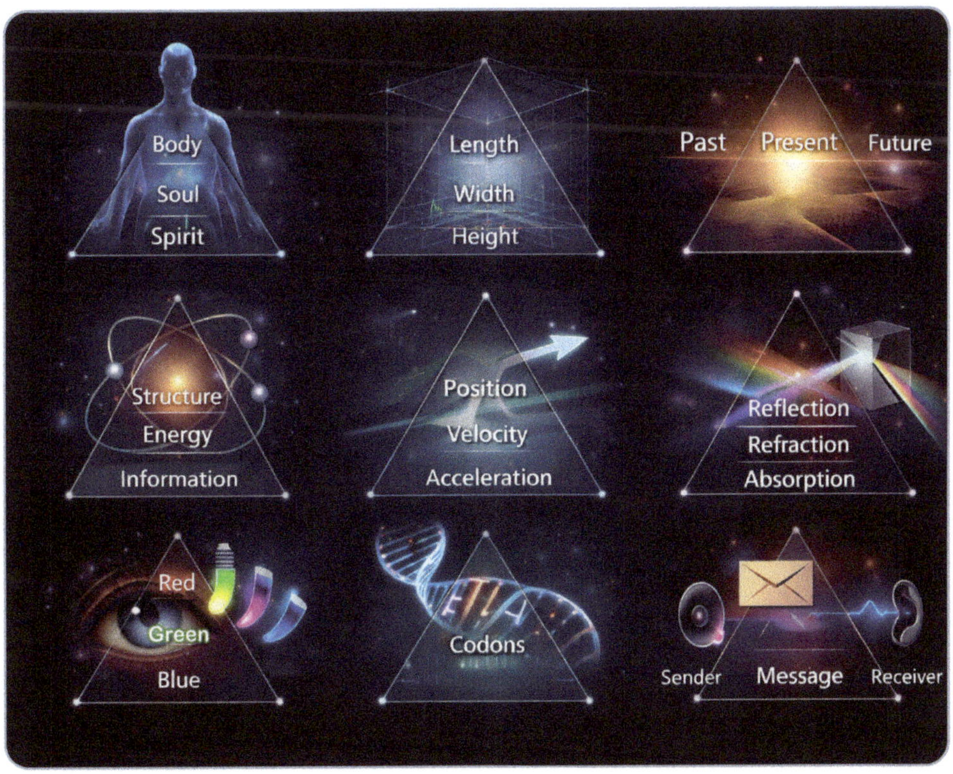

- Human beings are **body, soul, and spirit**.
- Space has **length, width, and height**.
- Time unfolds as **past, present, and future**.
- Matter itself displays **structure, energy, and information**.
- Motion in physics is described by **position, velocity, and acceleration**.
- Light interacts with matter through **reflection, refraction, and absorption**.
- Human vision perceives color through three primary receptors: **red, green, and blue**.
- The genetic code is written in **three-letter sequences (codons)**.
- Communication requires **a sender, a message, and a receiver**.

Light carries information, reveals truth, and communicates reality. It is a physical expression of God's deeper created order.

Again and again, one reality is expressed through three distinct but unified aspects. The triality of light resonates with this broader pattern. One reality expressed through three distinct but unified aspects.

Scripture reveals that God Himself is triune — God the Father, God the Son, and God the Holy Spirit. Therefore, it's no surprise that creation reflects its Creator.

Light, the first created reality in Genesis, may itself display this triune structure.

LIGHT, WORD, AND CREATION

The Bible consistently describes creation in terms of *speech* and *information*. God creates by *speaking*:

> *"Let there be light."* – ***Genesis 1:3***

Scripture repeatedly emphasizes the universe itself came into existence through God's word:

> *"By the word of the Lord the heavens were made ... by the breath of His mouth ... For He spoke, and it was done; He commanded, and it stood fast."* – ***Psalm 33:6-9***

The New Testament declares:

> *"In the beginning was the Word... All things were made through Him."* – ***John 1:1-3***

And we are told that:

> *"By faith we understand that the worlds were framed by the word of God, so that the things which are seen were not made of things which are visible."* – ***Hebrews 11:3***

God not only brought Creation into existence by divine speech — God the Son continually sustains it.

"upholding all things by the word of His power." – ***Hebrews 1:3***

The universe exists because of divine communication.

Creation itself is structured, ordered, and meaningful.

If reality is fundamentally informational, then the connection between light and word becomes profound. Light carries information, reveals truth, and communicates reality. It is a physical expression of the deeper created order.

The universe is not silent, it speaks:

"The heavens declare the glory of God." – ***Psalm 19:1***

TOWARD A DEEPER UNDERSTANDING

Viewing light as wave, particle, and information helps explain many mysteries:

- Why quantum behavior appears probabilistic
- Why particles display perfect identity
- Why observation affects outcomes
- Why the universe is mathematically ordered
- Why information plays a central role in physics

Light becomes more than energy. It becomes a bridge between matter and meaning.

Rather than reducing reality to blind mechanism, modern physics increasingly reveals structure, order, and intelligibility.

Light points beyond itself.

PREPARING THE WAY FOR QUANTUM CONNECTIONS

If light and matter are fundamentally informational, then the universe may be deeply interconnected at a level beyond classical space and time.

Quantum mechanics indeed reveals such connections. Particles separated by vast distances can remain linked in ways that defy classical expectations. Information appears to bind reality together.

This phenomenon, known as quantum entanglement, is one of the most astonishing discoveries in science.

In our next chapter we turn to this remarkable feature of light and the quantum world.

NOTICE AND DISCUSS

1. *What do you like about light being wave, particle, and information?*
2. *What features of light teach you about God's created world?*
3. *Comment on this: "A statement is not made of ink or sound waves. Meaning rides on physical carriers but is not identical to them. In the same way, physical processes may carry deeper informational content."*
4. *What did God put into light that you find most amazing?*

“

...upholding all things by the word of His power.

Hebrews 1:3

CHAPTER 4

Spooky Action at a Distance

QUANTUM LIGHT

Quantum entanglement is one of the most astonishing discoveries in modern science. It challenges our basic assumptions about reality, space, causation, and the nature of information itself.

When two particles become entangled, measuring one instantly determines the state of the other, even if they are separated by vast distances in space. The connection appears immediate, with no physical signal traveling between them.

Albert Einstein famously called this phenomenon **"spooky action at a distance,"** because it seemed to violate common-sense ideas about cause and effect. Yet decades of experiments have repeatedly confirmed that entanglement is real. Rather than a universe made of isolated objects, quantum physics reveals a deeply interconnected reality.

QUANTUM ENTANGLEMENT AND QUANTUM MECHANICS

The importance of quantum entanglement cannot be overstated. Physicist Erwin Schrödinger described it as "*the characteristic trait of quantum mechanics.*" In 2022, the Nobel Prize in Physics was awarded to three scientists (Alain Aspect, John Clauser, and Anton Zeilinger) whose experiments closed the final loopholes confirming that entanglement is a genuine feature of nature.

Their work demonstrated something remarkable: reality is not merely matter and energy interacting locally. Something deeper is at work.

UNDERSTANDING ENTANGLEMENT – A SIMPLE PICTURE

So what is quantum entanglement? From a bird's-eye view it can be pictured like this. Imagine two objects, A and B. Each can exist in one of two states, say red or blue. Through a subatomic interaction, object A becomes entangled with object B so that the two objects must now be described together as a single system.

One way to picture this is like two children on a playground who have decided they will always take opposite sides of every game. If child A chooses the swing, child B immediately heads for the slide. If child B grabs the swing first, child A runs to the slide. Their choices are always opposite.

In a similar way, entangled particles behave as a linked pair. When one particle is observed to be in one state, the other is immediately found in the corresponding opposite state. If particle A is red, particle B will be blue, and vice versa. Remarkably, this correlation holds no matter how far apart the particles are separated—even across the universe.

How do scientists know this works over long distances? In laboratory experiments, entangled particles are created together and then sent in opposite directions through optical fibers or across open air to detectors placed far apart. Researchers measure the particles independently and compare the results afterward. Again and again, the measurements show correlations that cannot be explained by ordinary local influences. Experiments have demonstrated this effect across kilometers on Earth and even between ground stations and orbiting satellites. The remarkable implication is that distance itself does not weaken the entangled connection.

Einstein could not accept this. As a committed materialist, he proposed that hidden variables must exist, such as pre-programmed instructions carried by the particles from the beginning. These hidden variables would explain their correlated behavior without requiring instantaneous influence.

But in 1964, physicist John Bell devised a test for Einstein's proposed hidden variables. Experiments repeatedly showed hidden variables do not exist. Later experiments eliminated remaining loopholes, overwhelmingly confirming that Einstein's explanation was incorrect.

One Nobel laureate, John Clauser, remarked: "I was very sad to see that my own experiment had proven Einstein wrong."

A UNIVERSE BOUND TOGETHER BY INFORMATION

What does entanglement imply?

Quantum entanglement suggests that reality is not merely a collection of independent objects. Instead, the universe appears fundamentally interconnected, bound together by relationships that transcend space. The October 2022 issue of *Quanta Magazine* described entanglement as a phenomenon where widely separated particles "appear to share information despite having no conceivable way of communicating." Physicist Terence Rudolph admitted that entanglement shows some things in quantum mechanics "transcend space-time."

This shifts attention to a profound question:

What is the fundamental nature of reality — matter, energy, or information?

Increasingly, physicists are suggesting that information itself may be the most fundamental layer of the universe.

MIT physicist Seth Lloyd, for example, has argued that 'the universe is a quantum computer,' a vast computational system processing information at every level. If reality is ultimately informational, it is not difficult to imagine the universe as reflecting ordered information sustained in the mind of God. Much like an embedded software system, perhaps

Reality is not merely a collection of independent objects. Instead, the universe appears fundamentally interconnected, bound together by relationships that transcend space.

God governs nature with precision through variables and pointers analogous to those in a computer language such as C, directing physical processes according to an underlying design.

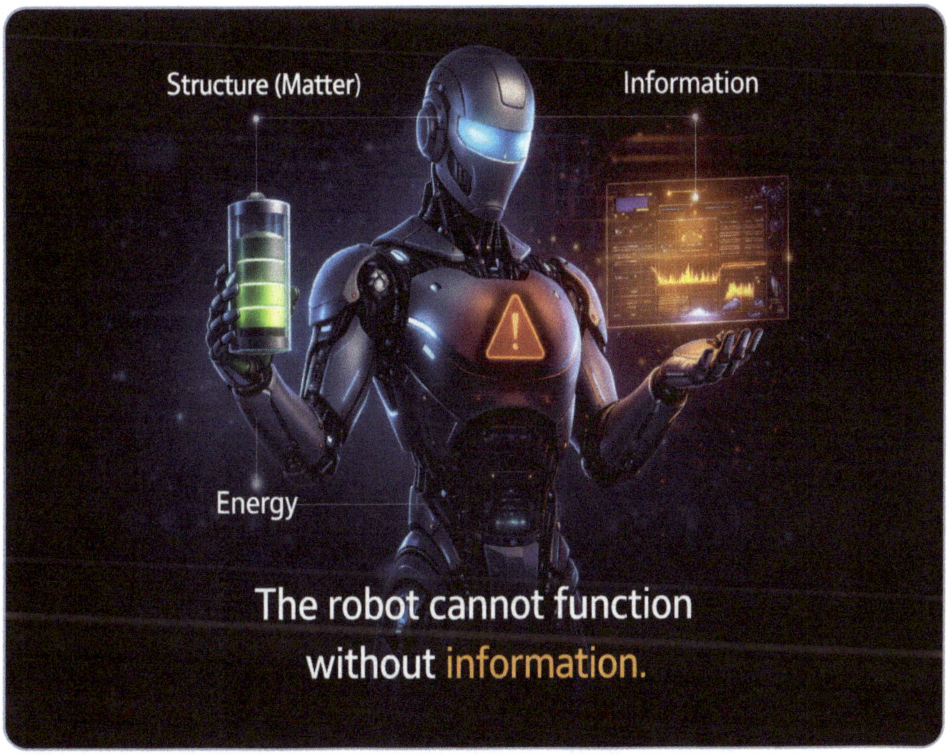

INFORMATION IS NOT MATTER OR ENERGY

Information is fundamentally different from physical substances. It has no mass. It occupies no space. It is not energy. As discussed in the previous chapter, information is not reducible to matter or energy. Quantum entanglement provides a striking physical example of this principle in action.

Yet information governs physical processes everywhere, from DNA to language to the laws of physics themselves. You cannot have meaningful information without intelligence. Information implies a source, a sender, and a message. If information truly lies at the foundation of reality, then the universe reflects not blind matter but ordered intelligence.

NONLOCAL REALITY AND THE LIMITS OF RELATIVITY

Quantum entanglement also challenges a long-standing principle of physics called **locality,** the idea that an object can only be influenced by its immediate surroundings.

Entangled particles contradict this principle. Their behavior is correlated instantaneously across space.

This raises another tension. Einstein's theory of special relativity posits that no information can travel faster than the speed of light. Yet entanglement appears to involve instantaneous correlations.

Physicists attempt to resolve this conflict in various ways. Some propose hypotheses like the "many worlds" interpretation, suggesting countless parallel universes, an idea many physicists themselves consider philosophically troubling and experimentally unfalsifiable. Others argue that no usable signal is transmitted, even though the correlations themselves are undeniable.

Despite these efforts, the mystery remains unresolved. Even mainstream physics admits that we don't fully understand how entanglement works without violating relativity. The March 2009 *Scientific American* described the phenomenon as presenting a "serious problem" for relativity.

The debate continues.

QUANTUM TUNNELING – ANOTHER MYSTERY

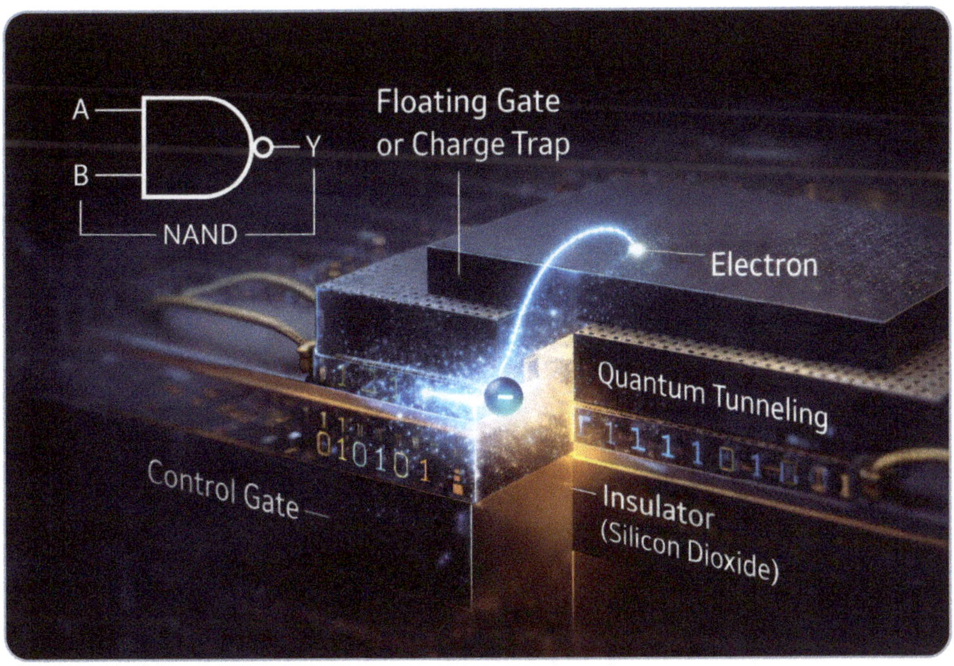

A related quantum phenomenon called **quantum tunneling** adds to the puzzle. In tunneling, particles sometimes pass through barriers that classical physics says they should not be able to cross. This effect even creates practical engineering challenges in modern computer memory chips. In NAND flash memory, for example, electron tunneling can cause tiny charge leaks over time, so engineers use error-correction and calibration techniques to keep stored data reliable.

Experiments measuring tunneling behavior have produced results that appear to occur faster than light. Some researchers have described aspects of tunneling as "robustly superluminal" meaning that the measured time through a barrier imply speeds exceeding the speed of light. While physicists debate how to interpret these findings, the quantum tunneling phenomenon highlights how quantum reality operates beyond classic material expectations.

Once again, physics encounters a world that appears to transcend and defy the classic limits of matter and energy.

God has created a wonderful world more expansive than we've imagined or comprehended so far:

> *"Now to Him who is able to do exceedingly abundantly above all that we ask or think, according to the power that works in us, to Him be glory in the church by Christ Jesus to all generations, forever and ever. Amen."*— **Ephesians 3:20-21**

Just think what discoveries await!

RESISTANCE AND SCIENTIFIC PARADIGMS

The history of quantum theory also reveals the human side of science. Early work on entanglement was often dismissed or treated as philosophical speculation. Some researchers were warned that pursuing such topics could damage their careers.

Theoretical physicist Max Planck once observed that scientific progress often advances not by persuasion but by generational change, "one funeral at a time". Paradigms can persist long after evidence begins to challenge them.

Quantum mechanics continues to provoke deep questions about the foundations of physics and the nature of reality.

Information implies a source, a sender, and a message. If information truly lies at the foundation of reality, then the universe reflects not blind matter but ordered intelligence.

A BIBLICAL PERSPECTIVE – CREATION SUSTAINED BY INFORMATION

From a biblical worldview, discoveries such as three properties of light, quantum entanglements, a universe bound together by information, and quantum tunneling are not surprising.

Scripture teaches that creation is upheld by divine command:

"He upholds all things by the word of His power." – ***Hebrews 1:3***

The universe is not self-sustaining. God's active governance continuously holds the universe together. If creation is sustained by divine information, then deep connections within reality should exist. Quantum coherence would not be an absurdity but a signature of design.

The apostle Paul likewise wrote:

"For since the creation of the world His invisible attributes are clearly seen, being understood by the things that are made, even His eternal power and Godhead..." – ***Romans 1:20***

Even the smallest particles testify to an amazing Creator.

LIGHT AND THE COMMUNICATION FABRIC OF CREATION

Light plays a central role in quantum phenomena. Photons can themselves become entangled, meaning they must be described as a single shared quantum system even when separated. In experiments, scientists often create entangled photon pairs and send them in opposite directions to distant detectors. When the photons are later measured, their results show correlations that cannot be explained by any local influences acting between the particles.

Light becomes part of the communication fabric of reality.

From the biblical declaration "*Let there be light*" to the quantum behavior of photons, light serves as a bridge between physical existence and informational order. Both of those lead to an intelligent and caring Designer.

Rather than chaos, quantum physics reveals coherence.

Rather than isolation, quantum physics reveals unity.

Rather than blind chance, quantum physics points toward design.

THE ULTIMATE INFORMATION GIVER

Reality is grounded in information that transcends matter and energy.

Quantum entanglement proclaims a profound message: reality is grounded in information that transcends matter and energy. This aligns naturally with the concept of God as the ultimate Information Giver — the One who speaks creation into existence and sustains it moment by moment (Genesis 1; Psalm 33:6-9; John 1:1-3; Hebrews 11:3, 1:3).

Spooky action at a distance does not undermine faith. Honest scientific discoveries point toward a universe upheld by divine intelligence, a creation in which even the smallest components reflect the wisdom and power of its Creator. As said above, the universe exists because of God's communication.

NOTICE AND DISCUSS

1. *What do you think Einstein meant by "spooky action at a distance?" (p. 44)*
2. *What picture in this chapter teaches you something you hadn't realized before and what is that new something?*
3. *Comment on this: "Quantum entanglement is one of the most astonishing discoveries in modern science. It challenges our basic assumptions about reality, space, causation, and the nature of information itself."*
4. *Now comment on this: "You cannot have meaningful information without intelligence. Information implies a source, a sender, and a message."*
5. *What would make us more readily open to truths that evidence reveals? Based on this from this chapter: "Paradigms can persist long after evidence begins to challenge them."*

What aspects of light's design reveal something meaningful to you about the Creator?

“

For by Him all things were created that are in heaven and that are on earth, visible and invisible

COLOSSIANS 1:16

CHAPTER 5

Light Reveals Cosmic Order

In 1990, NASA's COBE satellite detected a faint glow of microwave light filling the entire universe. This radiation, called the **Cosmic Microwave Background (CMB)**, appears in every direction. This CMB is a subtle background illumination permeating space itself.

Later missions such as the Wilkinson Microwave Anisotropy Probe (WMAP) and Planck mapped this glow in extraordinary detail. Scientists expected the radiation to be nearly uniform and random. Instead, they discovered subtle patterns and large-scale alignments within it, including a startling directional structure often called the

"Axis of Evil." The Axis of Evil refers to an unexpected pattern in the cosmic background radiation where temperature variations line up in a particular direction across the sky instead of being randomly spread out. The name itself reflects how troubling this discovery was for cosmologists attempting to explain the universe without a Creator or purposeful design.

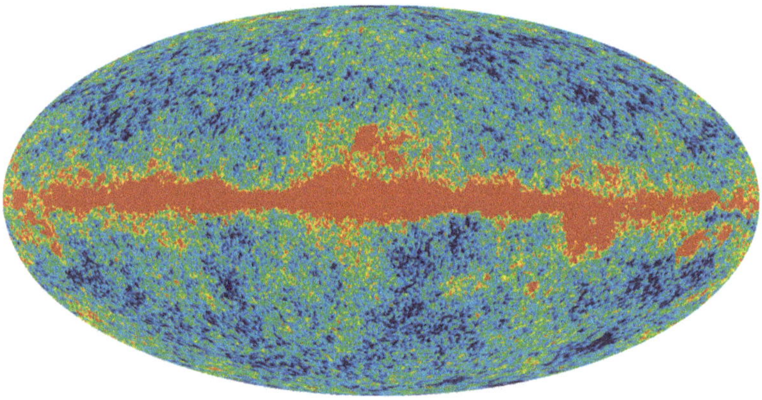

WMAP Full-sky Temperature Map, Q-Band

These findings surprised many researchers. If the universe is truly random and uniform on the largest scales, as many cosmological models assume, such large alignments should not exist. Some scientists even suggested that the data might point to a preferred direction in the universe or a special observational position. This idea conflicts with the Copernican Principle, which assumes that no place or direction in the universe is privileged.

But what if this patterning is not troubling at all? What if the universe is not random? What if the structure seen in cosmic light reflects intentional order, and our solar system, even our planet, indeed has a privileged place in the universe?

What if God created it? What if He created it specifically this way?

Some researchers have noted an intriguing possibility: the same conditions that allow life on Earth also allow discovery.

Our planet is uniquely positioned not only to exist, but to observe the universe with unusual clarity. Light reaches us in a way that makes the cosmos intelligible, a concept sometimes described as the "Privileged Planet."

A scientist doesn't have to believe in God to conclude this. Secular scientists Peter Ward and Donald Brownlee argue in their book *Rare Earth* that complex life depends on an extraordinarily rare combination of cosmic and planetary conditions. They underscore how uniquely suited Earth is for both life and scientific discovery.

> Earth is a privileged planet because of its unique position. We can observe the universe with unusual clarity, largely because of the light that reaches us and the way it reaches us.

LIGHT CARRIES STRUCTURE

The Cosmic Microwave Background (see page 47) is not merely radiation. This pervasive mass carries information. Tiny variations in temperature, differences of only millionths of a degree, reveal structure embedded in the fabric of the universe. These patterns form a map of cosmic organization on the largest observable scales.

Rather than chaos, the universe displays remarkable coherence.

> *"The heavens declare the glory of God."* – ***Psalm 19:1***

Even the faintest light in the universe appears to participate in this narrative.

LIGHT SHOWS A STRUCTURED UNIVERSE

As astronomical observations improved, scientists discovered matter itself is arranged in vast, organized structures spanning unimaginable distances. Galaxies are not scattered randomly. They form enormous clusters, filaments, and walls separated by vast cosmic voids. The universe resembles a complex and ordered network, sometimes described as a "cosmic web."

Among the largest known structures is the Sloan Great Wall, a massive formation of galaxies stretching across a vast region of space.

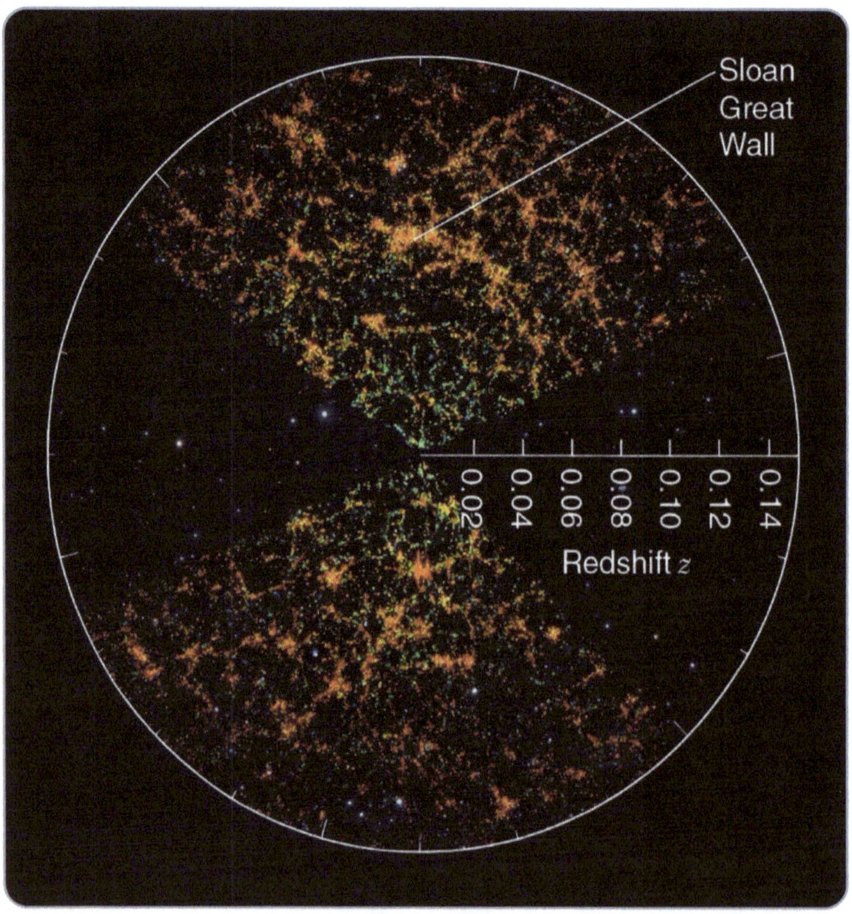

Through telescopes we see the architecture of the cosmos.

The measurements along the horizontal axis in the accompanying image represent **redshift (z)**, a value astronomers use to estimate the distance to these galaxies. Redshift occurs when light from distant objects is stretched toward longer, redder wavelengths (larger redshift values generally correspond to greater distances). Other enormous structures have since been discovered, forming filamentary patterns that span vast regions of the observable universe.

These structures raise profound questions. How did matter organize itself into such enormous patterns? How did structure arise on such immense scales?

Standard models propose that gravity, which is a *weak* force, gradually gathered matter over long periods of time. Yet the size, speed, and coherence of these structures continue to challenge simple gravitational explanations.

The universe appears organized in ways that suggest large-scale coordination rather than mere random accumulation, or results of weak forces.

Light reveals this structure to us. Through telescopes we see the architecture of the cosmos, not merely distant objects.

LIGHT SHOWS ORDER RATHER THAN CHAOS

The emerging picture of the universe is not one of disorder but of remarkable organization. For example:

- Radiation shows directional alignment.
- Matter forms vast coherent structures.
- Galaxies cluster into filaments and walls.
- Cosmic patterns extend across enormous distances.

The same light that reveals the tiniest atomic structures also reveals the largest cosmic architecture.

The universe appears ordered at every scale.

This order is not hidden. It is visible in light.

From quantum phenomena to galactic formations, reality exhibits:

- pattern,
- structure, and
- coherence.

Light exposes the universe as intelligible, something the human mind can study and understand.

LIGHT SHOWS A DESIGNED COSMOS THAT COMMUNICATES

For those who begin with a biblical worldview, cosmic order is not surprising. Scripture presents creation as purposeful, structured, and sustained by divine wisdom.

> *"He has established the world by His wisdom."* – ***Jeremiah 10:12***

For those who do not begin with a biblical worldview, the scientific evidence invites considering the truths of the Bible and considering God as Creator.

The structure revealed in cosmic radiation and galactic organization may therefore be understood not as accidental, but as evidence of intentional design.

The heavens do not merely exist. They communicate. They declare.

PLASMA PHYSICS AND ELECTROMAGNETISM

If the universe exhibits large-scale organization, an important question remains: what physical processes could produce such vast structure?

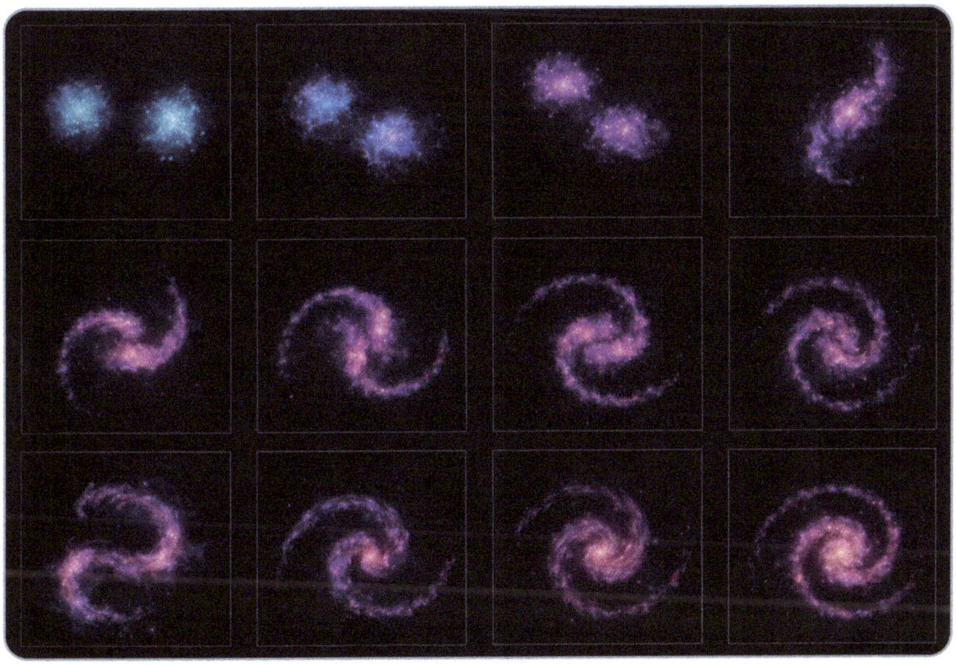

Plasma Lab simulation – 2 plasma filaments interact and create mini-galaxies in days, not millions of years.

Some researchers recognize that gravity alone cannot account for the observed cosmic architecture. Instead, they propose that electromagnetic forces and plasma behavior play a significant role in shaping large-scale structures.

Plasma— electrically charged matter— naturally forms filaments, currents, and network-like patterns. In a plasma universe, electrical currents flow through space along vast filamentary structures sometimes compared to cosmic "wires" that stars are connected to. Note that whenever you have an electric current (a flow of electric charges), a magnetic field is generated, linking electricity and magnetism as two aspects of the same electromagnetic interaction.

Laboratory experiments show that electrical processes in plasma can produce structures strikingly similar to those observed in space. Since plasma makes up most of the observable matter in the universe, electromagnetic forces could help organize matter on enormous scales.

Indeed, electromagnetic forces are orders of magnitude more powerful than the force of gravity.

Electrostatic Force vs. Gravity: A charged balloon pulls hair upward against Earth's pull, showing how electric forces can overpower gravity even on very small scales.

Even a small magnet can life these metal paperclips against earth's pull, showing how magnetic forces can easily overcome gravity.

As an example, a small magnet can easily lift a metal object against the pull of the entire Earth's gravity Likewise, a statically charged balloon or piece of cellophane can cling to a surface, while a charged comb or balloon can attract bits of paper, overcoming the downward force of gravity. In such cases, the gravitational attraction of an entire planet is defeated by a tiny electric charge or a simple magnetic field. These familiar demonstrations show how electromagnetic forces, even on very small scales, can readily overpower gravity, illustrating the enormous difference in strength between these fundamental forces.

The perspective of electromagnetic forces overpowering gravity leads naturally to a different view of cosmic energy and structure — one that emphasizes plasma physics and electromagnetic processes.

Do We Need Dark Matter?

Standard cosmology proposes the existence of unseen components called **dark matter** and **dark energy** to explain the motion of galaxies and the large-scale structure of the universe. For example, galaxies appear to rotate much faster than expected based on the gravity produced by their visible stars and gas. To explain this discrepancy, astronomers propose that large amounts of invisible matter must surround galaxies.

These unseen components are thought to make up most of the universe's mass and energy. However, dark matter and dark energy have never been directly observed in laboratory experiments. Their existence is inferred from gravitational effects and cosmological models.

If plasma cosmology is correct and large-scale cosmic structure is influenced by electromagnetic currents, then unseen matter and energy would not be necessary to explain galaxy formation and behavior. Why? Electromagnetic forces are vastly stronger than gravitational forces.

NOTICE AND DISCUSS

1. *Some large-scale patterns in the universe appear aligned across enormous distances. Why might these alignments challenge the idea that the universe is purely random?*
2. *The universe contains vast amounts of plasma – electrically charged gas. Why might electrical and magnetic forces be important in shaping cosmic structures?*
3. *The electromagnetic force is enormously stronger than gravity. How might this difference affect the way matter behaves in space?*
4. *Light allows astronomers to study distant objects across the universe. What kinds of information can scientists learn by studying the light from stars and galaxies?*
5. *When you look at images of galaxies, nebulae, or cosmic filaments, what features suggest order or structure to you?*

We now consider how a plasma universe may help explain one of the greatest puzzles about the cosmos.

"

The heavens
declare the
glory of God;
And the
firmament
shows His
handiwork.

PSALM 19:1

CHAPTER 6

What About Distant Starlight?

Modern research suggests space is not truly empty. Even what we call vacuum has structure and activity. At the same time, much of the visible matter in the universe exists not as solid matter but as plasma. As described in chapter 5, plasma is an electrically active state of matter composed of charged particles. These charged particles respond strongly to electromagnetic force. Plasma often exhibits filamentary behavior. This means plasma forms long, thread-like strands that weave through space like glowing cosmic wires. Plasma would thus be a very strong candidate to explain the vast filamentary structures throughout space. In this electrically dynamic environment energy flows through large-scale currents and fields. Stars, nebulae, and galaxies exist here.

Rather than moving through a truly empty void, light travels through a universe with structure. In many regions, space contains plasma, fields, and energetic activity rather than simple emptiness. In addition, modern physics suggests that even a vacuum is not entirely empty. This means light may travel through a far more complex environment than people once imagined.

James Webb Space Telescope (JWST) Deep Field — Thousands of distant galaxies in this tiny patch of the universe raise questions about how their light has reached Earth.

Science has already demonstrated that light travels more slowly in materials such as glass, water, air, and plasma. Interactions with atoms and charged particles affect light's transmission, direction, and apparent speed. So, a universe filled with plasma and electromagnetic activity would allow light to move differently than in a simple empty space.

This possibility raises an important question: if the structure of the cosmos has changed over time, could light have traveled much more rapidly under different conditions?

One of the most frequently asked questions about the cosmos concerns distant starlight. If stars and galaxies are extremely far away, how has their light reached Earth? This question often assumes the universe has always operated exactly as it does today: the same structure of space, same physical conditions, and same behavior of light throughout all of history.

But the Bible describes creation in a process purposefully structured by God. Could conditions present during creation have differed significantly from those we observe now?

> Light is central to understanding creation and the universe.

Rather than assume an unchanging universe, let's ask a different question: What were the original conditions of creation when God first spoke light into existence?

Was the structure of space, the properties of matter and energy, or the behavior of light different at the beginning? If so, then light could have traversed vast distances far more efficiently than under present conditions.

Once again, light itself becomes central to understanding creation.

LIGHT CREATED BEFORE THE STARS

The biblical account of creation provides an important starting point.

According to Genesis, God introduced light on the first day. The sun, moon, and stars were not appointed until the fourth day. That is, light existed prior to the heavenly bodies that now bear it.

The source of light is not the stars themselves but the command of God.

The properties of light were established by divine design before the structure of the heavens was fully arranged.

If God created the universe in a mature and functional state, then its physical systems may have operated under conditions different from those observed today. The laws of physics remained orderly, but the conditions of creation, including the structure of space and the energy density of the vacuum, may have differed from those we observe now.

This ordered creation would have been different from an eternal, self-existing system. This means the universe did not arise gradually from an eternal system of matter and energy but instead began as a purposeful creation with built-in order.

Understanding distant starlight therefore requires understanding the nature of creation itself.

A UNIVERSE FILLED WITH PLASMA

In a plasma cosmos, also called a plasma universe:

- Plasma is widespread throughout the visible universe
- Electromagnetic forces play a central role
- Plasma often forms filaments, currents, and large-scale structures

More broadly, modern physics also suggests that:

- Space is not a featureless emptiness
- The vacuum has measurable properties
- Light does not travel through a simple void

If the structure of this medium differed during creation, light may have traveled differently than it does today. A more uniform or differently organized cosmic environment could have allowed light to move across enormous distances with far greater efficiency.

This perspective reinforces a central theme of this book: light is deeply connected to the structure of reality.

The Casimir effect provides evidence that empty space is not truly empty. When two metal plates are placed extremely close together in a vacuum, they experience a tiny force pushing them together, suggesting that the vacuum itself has measurable structure.

LIGHT, SPACE, AND ZERO-POINT ENERGY

Modern physics has revealed something remarkable about space. Even in a perfect vacuum, where all matter and radiation are removed, measurable energy remains. This ever-present background activity is known as zero-point energy (ZPE).

Zero-point energy arises from constant electromagnetic fluctuations that persist even at the lowest possible temperatures. Because of zero-point energy, space itself possesses:

- structure,
- energy, and
- activity

even in a vacuum. A vacuum is not a passive void but an active physical medium that influences atomic behavior, the propagation of light, and the structure of matter itself.

This newly discovered conclusion is widely accepted within modern physics. Because light travels through space, the properties of this underlying vacuum directly affect how light behaves.

LIGHT AND THE BIBLICAL DESCRIPTION OF A STRETCHED UNIVERSE

Scripture repeatedly describes God as the One who "stretched out the heavens" (Isaiah 40:22; Jeremiah 10:12; Zechariah 12:1). While the Bible is not a physics textbook, this language describes a dynamic creation in which the fabric of the cosmos itself was extended or expanded during creation week.

If the universe was initially created with little or no vacuum energy and was then "stretched," the energy of space would increase as the fabric of the universe expanded, much like tension increases when a rubber band is stretched.

As the structure of space changed, the electromagnetic properties of the vacuum would also change.

Because the speed of light depends on the electric and magnetic properties of space, increasing vacuum energy would naturally influence how light travels. As the vacuum increases, light could travel less rapidly than under earlier cosmic conditions. Light could have gradually slowed as the structure of space developed.

This possibility suggests that the key question is not merely how far light travels, but what kind of universe it travels through.

LIGHT AND EVIDENCE FROM QUANTIZED REDSHIFT

An intriguing line of research may provide observational support for the idea that space has large-scale structure affecting the behavior of light.

Some astronomers, including William Tifft and Halton Arp, have reported that galactic redshifts (*the stretching of light toward longer wavelengths*) appear in intervals rather than forming a smooth continuum. These patterns are known as **quantized redshifts** (see illustration on page 55).

Rather than interpreting redshift solely as a measure of distance or velocity, some researchers propose that these discrete intervals reflect structured properties of space itself that influence light during its journey.

One proposed explanation connects these redshift intervals to the quantized nature of atomic energy levels. In atoms, electrons do not occupy continuous ranges of energy but exist only in discrete orbitals separated by specific energy gaps. Transitions between these levels produce precise electromagnetic effects.

Remarkably, the predicted values derived from these calculations, called a quantum framework, match observed redshift intervals reported by astronomers Tifft and Arp. Calculated "redshift shells" correspond to discrete velocity values such as 2.667, 5.33, 8.00, 16.0, 24.0, 72.0, 144, 216, and 288 km/s— the same values reported observationally.

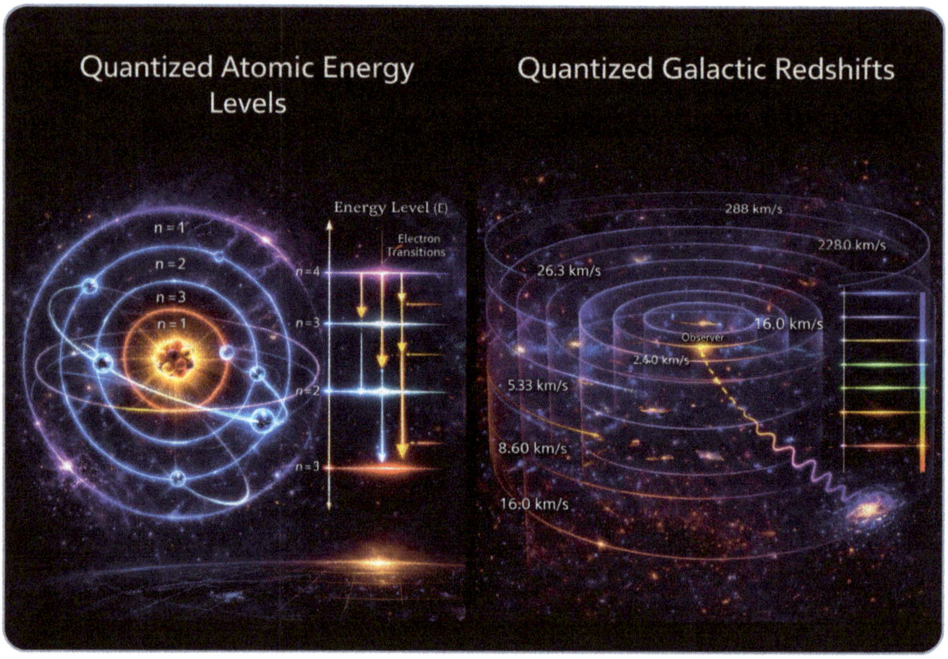

From Atoms to the Cosmos – Just as electrons occupy fixed energy levels rather than a continuous range, some researchers propose that galactic redshifts reflect quantized structure in space itself.

> Creation scientists, including physicist D. Russell Humphreys, suggest that quantized redshift patterns may reflect a large-scale structure of the universe centered on our galaxy, consistent with a created cosmos in which humanity occupies a special position.

If further confirmed, such findings aligning electron orbital intervals with cosmic redshift intervals would be difficult to explain as coincidence.

These findings could provide a possible physical framework consistent with the biblical description of God "stretching out the heavens" (Isaiah 40:22; Jeremiah 10:12; Zechariah 12:1). This framework offers an alternative explanation for observations often interpreted as evidence of billions of light-years of distance.

CHANGING PHYSICAL CONDITIONS AND THE SPEED OF LIGHT

Physicist Barry Setterfield has proposed a model linking the behavior of light to changes in the energy structure of the vacuum.

In Setterfield's model, zero-point energy underlies the electric and magnetic properties of space, which in turn influence atomic processes, clock rates, and the speed of light. Reread the previous section for more details. If the background energy of space increased as the universe developed, consistent with a "stretching" of the heavens, multiple physical quantities would change simultaneously:

- atomic energy levels
- rates of physical processes
- the ticking of atomic clocks
- the speed at which light propagates

EPOLA - A Structured Medium of Space?

Physicist **Yaakov Simhony** proposed that space may have an underlying structure rather than being truly empty. His **EPOLA model** (E.lectron-Po.sitron La.ttice) suggests the vacuum could consist of an ordered lattice of electron-positron pairs.

This idea builds on Nobel Prize-winning physicist **Carl David Anderson's** discovery of the positron, showing that particle pairs can emerge from what appears to be empty space.

In Simhony's model, the vacuum behaves as an electromagnetic medium.

If such a structure exists, properties of this lattice could influence how light travels through space. For example, variations in the density or arrangement of the medium could affect light's movement and speed.

Even modest differences in these conditions would dramatically affect how quickly light traverses cosmic distances.

Under earlier conditions with lower vacuum energy, light could have traveled far more rapidly than it does today.

This framework does not require nearby stars or vast ages. Instead, it suggests that the physical conditions governing light may have changed as the structure of the cosmos developed.

LIGHT, TIME, AND PHYSICAL MEASUREMENT

> The immense ages often assigned to the universe may reflect changing physical processes—not necessarily the actual passage of time.

Because measurements of time depend on atomic processes, any change in the vacuum's properties would also affect how time is *measured*; that is, the clocks used to measure it. If physical processes operated differently during creation, then the flow of time as measured by atomic systems may have differed from present conditions.

What appear today as immense cosmic timescales may reflect changes in physical processes of the universe rather than long durations. Time measurements depend on physical systems, and those systems may themselves be influenced by the structure and energy of space.

HAS THE SPEED OF LIGHT ALWAYS BEEN CONSTANT?

Modern physics generally assumes the fundamental constants of nature, including the speed of light, have always remained fixed. Yet if physical processes depend on the structure and energy of space, and if the properties of the vacuum have changed as the universe developed, then the behavior of light may not have always been the same.

Rupert Sheldrake Challenges Scientific Orthodoxy in His TED Talk "The Science Delusion." This talk was later removed from TED's platform under pressure from secular critics.

Cambridge biochemist Rupert Sheldrake has drawn attention to historical measurements of the speed of light that suggest the possibility of change.

After examining older physics handbooks and experimental records, he found that the measured value of the speed of light appeared to decline between approximately 1928 and 1945 by about twenty kilometers per second. Laboratories around the world reported progressively lower values during this period, despite extremely small reported experimental errors. After 1948, the measured values rose again and then stabilized.

Intrigued by this pattern, Sheldrake sought clarification from the United Kingdom's National Physical Laboratory (NPL) in Teddington, one of the world's leading authorities on precision measurement. He reports discussing the issue with the head of metrology, who responded:

"You've uncovered the most embarrassing episode in the history of our science."

When Sheldrake asked whether the speed of light itself might have changed, the official rejected the idea and instead attributed the earlier variations to what he described as **"intellectual phase locking."** This term refers to the tendency of scientific measurements to converge toward *expected* values. Researchers have been known to adjust methods or interpretations, so results align with accepted theory or with the results of others.

When Sheldrake asked how modern science avoids the same problem today, he was told the issue had effectively been resolved by redefining the speed of light itself. Beginning in 1972 and formally adopted in 1983, the International System of Units fixed the speed of light at exactly 299,792,458 meters per second and defined the meter in terms of light's travel.

Under this system, the speed of light is **fixed by definition rather than determined by experiment**, and it now defines the standard of length.

Because time is measured by atomic clocks, and atomic processes depend on the same underlying physical properties, the measuring system and the measured quantity become linked.

If fundamental constants had changed together, the units used to measure them would change as well, making any variation effectively undetectable.

Thus, the decision beginning in 1972 and formally adopted in 1983 determined that the speed of light must appear constant by definition. Rather than testing whether the speed of light changes or has changed, the system defined it in a way that prevents change from ever being detected.

This approach is reminiscent of how evolution was redefined as mere small-scale change, making it automatically true by definition. At Real Science Radio, we call this "**intellectual phase-prestidigitation,**" a kind of scientific sleight of hand that locks in the answer before the experiment even begins.

> Be wary of approaches that lock in the answer before the experiment even begins.

LIGHT AND QUESTIONS ABOUT MEASUREMENT FOUNDATIONS

Concerns about the interpretation of measurement standards have also been raised by physicist Louis Essen, a senior scientist at the National Physical Laboratory and developer of the cesium atomic clock, the foundation of modern precision timekeeping.

Essen played a central role in establishing modern frequency standards and helped develop the cesium atomic clock technology that defines the modern second. Yet later he questioned prevailing interpretations of atomic clock experiments, particularly their use in supporting Einstein's theory of relativity. This theory proposes that time itself can vary depending on motion and gravity, so clocks moving under different conditions will measure different amounts of time.

Essen emphasized that atomic clocks measure specific physical processes and cautioned against assuming that such measurements automatically settle deeper questions about the nature of time or physical reality.

Essen's work highlights the importance of carefully distinguishing between physical phenomena and the methods used to measure them.

This raises a broader principle: measurements depend on physical processes. If those processes are affected by changing conditions in space, then the measurements themselves may reflect those changes.

What is the nature of time? Of physical reality?

A UNIVERSE SUSTAINED BY ENERGY

The discovery that space contains sustaining energy also resonates with the biblical description of a creation upheld by an ongoing source of power—God Himself— rather than a self-existing mechanism.

The Bible declares God is:

"...upholding all things by the word of His power." – ***Hebrews 1:3***

From the Bible's perspective and the evidence of discoveries such as quantum entanglement, large-scale patterns in the cosmic microwave background, and plasma filaments stretching across space, the energy underlying the universe may reflect the sustaining order built into creation.

LIGHT AS A MESSENGER OF CREATION

Throughout this book we have seen that light carries information. It communicates the composition, structure, and motion of distant objects. The universe speaks through light.

"Day unto day utters speech, And night unto night reveals knowledge."
– ***Psalm 19:2***

For the heavens to declare anything, their message must be received. You and I are some of these receivers. Scientific study verifies light's communication.

Light serves as the messenger carrying this testimony of God's glory across creation. If the universe was designed to communicate information, then the transmission of light across vast distances is not surprising. Creation may be structured in a way that ensures its message is seen.

EARTH — AN IDEAL PLATFORM FOR OBSERVING THE UNIVERSE

Our ability to see distant starlight depends not only on the stars themselves but also on where we are observing from. Remarkably, Earth is extraordinarily well suited for viewing the cosmos. Several features of our planet make astronomical observation possible.

> Earth is not only a home for life—it is also an observatory from which the universe can be seen.

- **A transparent atmosphere.** Earth's atmosphere allows visible light to pass through while blocking harmful radiation. This protective filter lets us see the stars clearly while shielding life from dangerous cosmic energy.

- **A stable rotating platform.** Earth's rotation slowly sweeps the sky overhead each night, allowing observers to view the entire celestial sphere over time. This natural motion made systematic astronomy possible throughout human history.

- **A clear optical window.** The wavelengths most useful for studying stars and galaxies pass directly through Earth's atmosphere. These optical "windows" allow astronomers to detect the composition, temperature, and motion of distant objects through spectroscopy.

- **A location within a transparent region of the Milky Way.** Our solar system lies in a relatively clear part of our galaxy, away from dense dust clouds that would otherwise obscure our view of the broader universe.

Because of these features, Earth provides an exceptional vantage point for observing the heavens. The same conditions that allow life to exist here also allow us to study the cosmos.

From a biblical perspective, this is not surprising. If the heavens were created to declare the glory of God (Psalm 19:1-2), then the place where human beings live would naturally be well suited to receive that message.

A CREATION DESIGNED TO BE SEEN

From a biblical perspective, the distant starlight question (recall chapter 6) becomes an invitation to consider the nature of creation itself. God introduced light before the stars. The properties of light were determined by divine command, not by blind physical processes. If the universe was designed to reveal His glory, its structure, laws, and physical conditions would naturally support that purpose.

The distant stars still shine.

The stars' light still arrives.

The stars' message still declares.

And once again, light points beyond itself to its Creator.

NOTICE AND DISCUSS

1. *Why do some scientists believe distant starlight raises questions about the age of the universe?*
2. *Genesis describes light existing before the Sun and stars. What might this reveal about the nature of light?*
3. *Some astronomical observations raise questions about how light travels across enormous distances. What explanations have scientists proposed?*
4. *Observations such as quantized redshift suggest that the universe may be more structured than scientists once believed. Why might this surprise researchers?*
5. *Some researchers have suggested that physical conditions in space may have changed over time. Why would this matter when interpreting distant starlight, and what might it reveal about the design of the universe?*

In our next chapter we turn to a different perspective on our nearest star – the Sun.

“

Then God made two great lights: the greater light to rule the day, and the lesser light to rule the night.

GENESIS 1:16

CHAPTER 7

Light and The Electric Sun

A REAL SCIENCE RADIO PERSPECTIVE

In the previous chapters we saw that the universe appears

1. **highly structured,**
2. **electrically active, and**
3. **filled with plasma and ZPE rather than empty space.**

If electromagnetic processes play a major role in shaping cosmic structure, an important question follows: might these same processes also influence our nearest star, the Sun?

The Sun – A Dynamic Plasma Environment

Scientists studying the Sun face many of the same limitations encountered in cosmology (study of the universe) where direct experimentation is impossible. Observations must be interpreted from an enormous distance, instruments must withstand intense heat and radiation, and the Sun's extreme brightness creates significant visual and measurement challenges. Because researchers cannot directly sample the Sun's interior, much of what is known comes from indirect observations such as spectroscopy and computer modeling. As a result, conclusions often rely heavily on inference and underlying assumptions.

THE STANDARD MODEL OF THE SUN

Most textbooks teach that the Sun shines because of nuclear fusion deep in its core. In this standard model, hydrogen atoms compress under immense pressure and temperature and then release energy that eventually emerges as heat and light.

But how well does this line up with the evidence we have available to us?

Science educator **Ellen McHenry**, in her book *Sun Puzzles*, highlights six puzzling observations about the Sun that remain difficult to explain within a strictly nuclear- fusion-only framework. She listed a number of solar mysteries —detailed in the next several paragraphs— that are more naturally explained if **plasma physics** and large-scale electrical processes play a major role in how the Sun operates. You'll recall that plasma physics is the study of electrically charged gases whose particles respond strongly to electromagnetic forces. Because the Sun is made almost entirely of plasma, these electrical effects may play an important role in how it behaves.

This alternative view emphasizes that space is not empty but filled with plasma as discussed in earlier chapters, electrically charged matter that responds strongly to electromagnetic forces.

OBSERVATIONAL PUZZLES OF THE SUN

Here are some of the observational puzzles Ellen McHenry identified. These puzzles about the Sun challenge simple expectations of an internally-heated nuclear object.

Hot core to hotter surface is one of solar physics' most enduring questions.

The Hot Corona Problem

The Sun's outer atmosphere, the corona, reaches temperatures of over one million degrees Celsius, far hotter than the visible surface. If heat originates primarily in the core, temperatures would be expected to decrease with distance from that source. Instead, the outermost layer is dramatically hotter.

This reversal remains one of solar physics' most enduring questions.

Heavy Elements in the Solar Atmosphere

The Sun's atmosphere is extremely thin and composed mostly of lightweight hydrogen plasma. Yet scientists detect heavier elements such as iron there. Why would heavy elements remain suspended in such a lightweight and tenuous environment rather than falling inward? Researchers acknowledge that the mechanism responsible for this selective transport is not fully understood.

One possible explanation is electrical processes in plasma naturally separate and move charged particles (#2 and #3 in the page 64 list).

Cool Sunspots

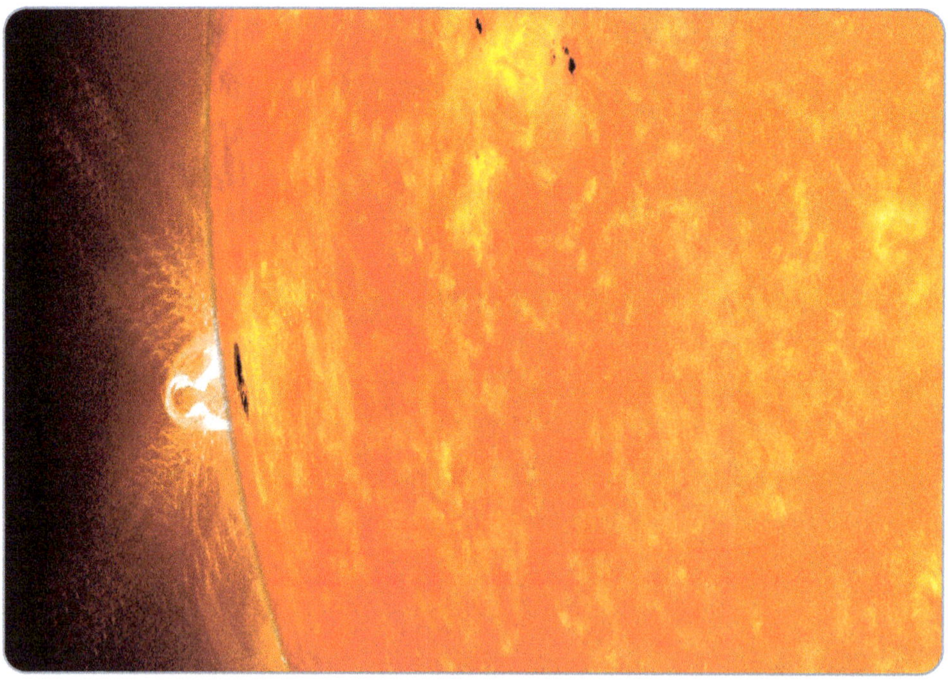

Sunspots are the dark, cooler regions on the Sun's surface. NASA, 2014.

Sunspots appear as dark regions on the Sun's surface and are cooler than their surroundings. If heat flows outward from the Sun's core, openings in the Sun's surface might be expected to reveal hotter regions, not cooler ones.

Instead, sunspots behave more like interruptions in energy flow than like vents releasing internal heat. Some researchers interpret sunspots as evidence of complex electromagnetic activity.

The Solar Wind

The Sun continually emits streams of charged particles traveling at extraordinary speeds. Several features of the solar wind remain puzzling:

- Solar winds are both fast and slow.
- The particles accelerate as they move away from the Sun.
- Solar wind has even been observed to nearly stop for short periods.

Electrical forces can accelerate charged particles across large distances, providing a possible framework for understanding these behaviors (#1 and #2 of the page 64 list).

A PLASMA UNIVERSE AND ELECTRICAL PROCESSES

As discussed in earlier chapters, modern observations show that most visible matter in the universe exists in the plasma state, a mixture of charged particles such as electrons and ions (atoms that have lost or gained electrons and therefore carry an electric charge). If the universe is filled with electrically responsive plasma and ZPE, then stars like our Sun may not function as completely isolated objects. Instead, they may interact with their surrounding environment through large-scale electromagnetic processes, consistent with reports of electric currents associated with astrophysical jets and other cosmic structures.

McHenry traced the work of several pioneering scientists whose research helped establish plasma physics and electromagnetic processes in space:

- **Kristian Birkeland** studied electrical currents producing auroras.
- **Hannes Alfvén**, Nobel Prize–winning physicist who developed plasma physics.
- **Irving Langmuir** identified charged layers within plasma.
- **Ralph Juergens** proposed the Electric Sun hypothesis.

Their work suggests that electrical currents may operate throughout space. These electrical currents influence stars, planets, and galaxies. Electrical processes naturally accelerate charged particles, separate elements, and generate complex energy flows. These properties provide a possible framework for understanding several of McHenry's observations and other phenomena observed within our solar system.

THE ELECTRIC SUN MODEL

In the Electric Sun model, the Sun is viewed as part of a larger electrical system in which:space contains charged particles,

- electrical currents flow across cosmic distances,
- and the Sun interacts with this electrical environment.

Within this framework, the Sun can be viewed somewhat like an electrode within a larger electrical circuit. Incoming charged particles could help drive phenomena such as solar winds, flares, and the complex loops of glowing plasma observed in the Sun's atmosphere.

The electric sun model highlights the possibility that electromagnetic forces play a greater role in cosmic structure than commonly assumed. Yet even these physical explanations raise a deeper question about the nature and origin of light itself.

LIGHT BEFORE MECHANISM

Scripture presents light as foundational to creation:

"God said, 'Let there be light.'" – **Genesis 1:3**

> The study of the Sun is more than a study of physics—it is a study of how God sustains light within creation.

Light appeared before the Sun and stars. The heavenly bodies later serve as bearers of light. The heavenly bodies are not light's ultimate source.

Whether the Sun's energy comes primarily from internal fusion, external electrical processes, or some combination of mechanisms, the biblical order remains striking: light is primary.

The study of the Sun therefore becomes more than a study of physics. It becomes a study of how God sustains light within creation.

THE SUN AND THE GREATER LIGHT

The Bible ultimately points beyond the physical Sun:

"...the LORD will be to you an everlasting light." – **Isaiah 60:19**

The Sun itself may be only a temporary instrument through which divine light is expressed. The Sun's mysteries remind us the physical universe reflects deeper realities.

If light governs the structure of the cosmos, does it also shape living systems such as people? We now turn from the cosmos to life itself.

NOTICE AND DISCUSS

1. *Scientists have discovered several surprising features of the Sun, such as the extremely hot corona and the presence of heavy elements in the solar atmosphere. Why do these observations puzzle researchers?*
2. *Most of the visible universe is made of plasma, an electrically charged state of matter. How might electrical processes influence what happens in stars like the Sun?*
3. *The Sun continually emits a stream of charged particles called the solar wind. What does this tell us about the Sun and its interaction with the surrounding space environment?*
4. *Light from the Sun allows scientists to study its composition and behavior. What kinds of information can astronomers learn from sunlight?*
5. *What might the complex behavior of the Sun suggest about the design and order of the universe?*

"

The Spirit
of God has
made me,
And the
breath of
the Almighty
gives me life.

Job 33:4

CHAPTER 8

Light in Biology is the Spark of Life

Light is not only cosmic — it is biological. Every living cell maintains an electrical potential across its membrane. Your heartbeat, your brain activity, and your muscle movements all depend on finely tuned electrical signals. Life operates through controlled flows of energy and information, coordinated with astonishing precision.

At the very moment of conception, when sperm meets egg and a new life begins, a measurable voltage shift occurs. Many describe this as a "spark of life."

This rapid electrical change triggers biochemical cascades that begin human development. A new organism is initiated not by random chemistry, but by a sudden transition into ordered electrical activity.

This initial electrical spark is not merely symbolic; it serves a critical protective function. When a sperm successfully fertilizes an egg, the egg's membrane undergoes an immediate electrical change called membrane depolarization. After this comes a rapid wave of calcium ions spreading across the cell. This precisely timed response prevents additional sperm from entering the egg.

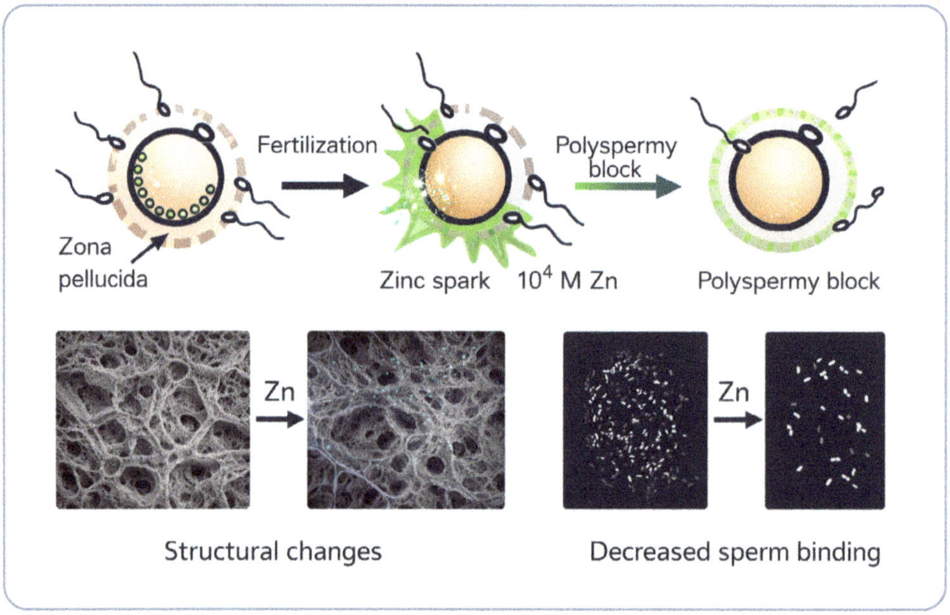

When a sperm fertilizes an egg, a burst of zinc and a rapid electrical change help block additional sperm from entering. Based on published scientific research.

Without this protective electrical mechanism, a condition called polyspermy would occur, multiple sperm fertilizing a single egg. The new human life would not survive.

To prevent this and protect the new life, the egg rapidly activates multiple coordinated defenses:

- An electrical shift blocks additional sperm entry within seconds.
- A biochemical reaction hardens the outer layer of the egg.
- Cellular signaling reorganizes the egg's surface receptors.

This system operates with remarkable precision and timing. A delay of a fraction of time can lead to the end of the new life. Life's beginning therefore depends on a finely tuned sequence of electrical and biochemical events that preserve order from the very first instant of human development.

> Life reflects both
> - energy and
> - meaning
>
> woven together in design.

The spark of life is not merely metaphorical. It is measurable, structured, purposeful, and protective—a testimony that life reflects both energy and meaning woven together in extraordinary design.

Life begins at conception, and in that very moment a measurable electrical spark announces a new human life.

LIFE RUNS ON ELECTRICAL ENERGY

All living systems are fundamentally electrical.

Cells communicate through voltage gradients.

Tissues form through bioelectric signaling.

Even wound healing depends on electrical currents guiding cellular behavior.

Research in developmental biology has shown that electrical fields help determine the shape and structure of organisms long before full genetic expression occurs.

Inside every cell, tiny molecular machines maintain electrical gradients across membranes. Mitochondria generate energy by moving charged particles across microscopic barriers, creating voltage differences that power the synthesis of ATP, the universal energy currency of life.

At its deepest level, life is sustained by controlled electron flow. Living organisms are not merely chemical systems, they are dynamic electrical systems. Recent research suggests that even water inside living cells may store electrical charge and energy from light.

LIGHT WITHIN CELLS

Biophysicist Gerald Pollack has proposed that water inside living cells forms a structured state known as the "fourth phase of water," or exclusion-zone (EZ) water. Unlike ordinary liquid water, this structured form carries electrical charge and can store energy from light.

Experiments show that light can separate electrical charge in water, generating measurable voltage. Because living cells are largely composed of water, this process may contribute to the electrical properties of life.

If confirmed, these findings suggest that light not only sustains life externally but may help organize energy and structure within living cells.

LIFE EMITS LIGHT

Remarkably, living organisms also produce light.

Sensitive instruments have shown that plants, animals, and even human beings emit extremely faint light known as **biophotons**. These ultraweak photon emissions arise from metabolic processes within cells and appear to reflect the state of living tissues. The human body literally glows, though the light is far too weak for the eye to detect.

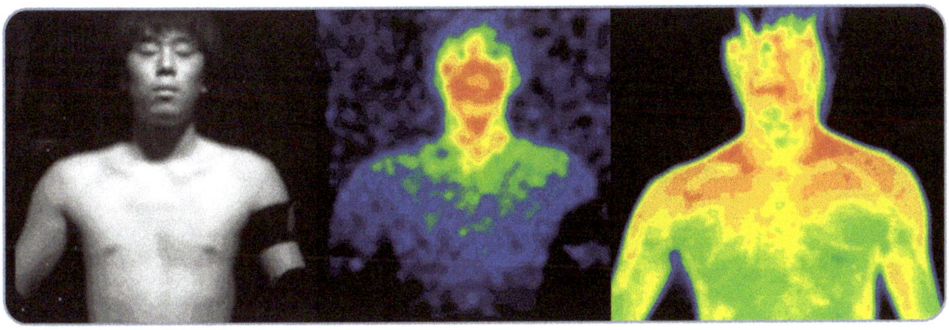

Ultraweak photon emission from a human body ("biophoton"). Adapted from Kobayashi et al., PLOS ONE (2009), CC BY.

When life ends and cellular activity stops, the emission of this light ceases.

Some researchers suggest that these light emissions may even play a role in cellular communication and regulation. Whether or not this proves true, the phenomenon itself is striking: life is not only powered by light — it radiates light.

PHOTOSYNTHESIS – CAPTURING LIGHT TO SUSTAIN LIFE

The connection between light and life is most clearly seen in plant photosynthesis, which depends specifically on visible light.

Plants capture photons of light within this range to convert light energy into chemical energy, forming the base of nearly every food chain on Earth. The oxygen we breathe is itself a product of this light-driven process.

If the Sun emitted primarily ultraviolet radiation, life would be damaged by constant molecular destruction.

If the Sun emitted mostly infrared radiation, the energy would be insufficient to drive the chemistry of life.

But instead, the Sun provides the visible spectrum that lies precisely within the range that supports biological function. In this sense, nearly all life on Earth runs on captured sunlight. The energy of the sun:

- is transformed into sugars,
- fuels metabolism, and
- ultimately sustains entire ecosystems.

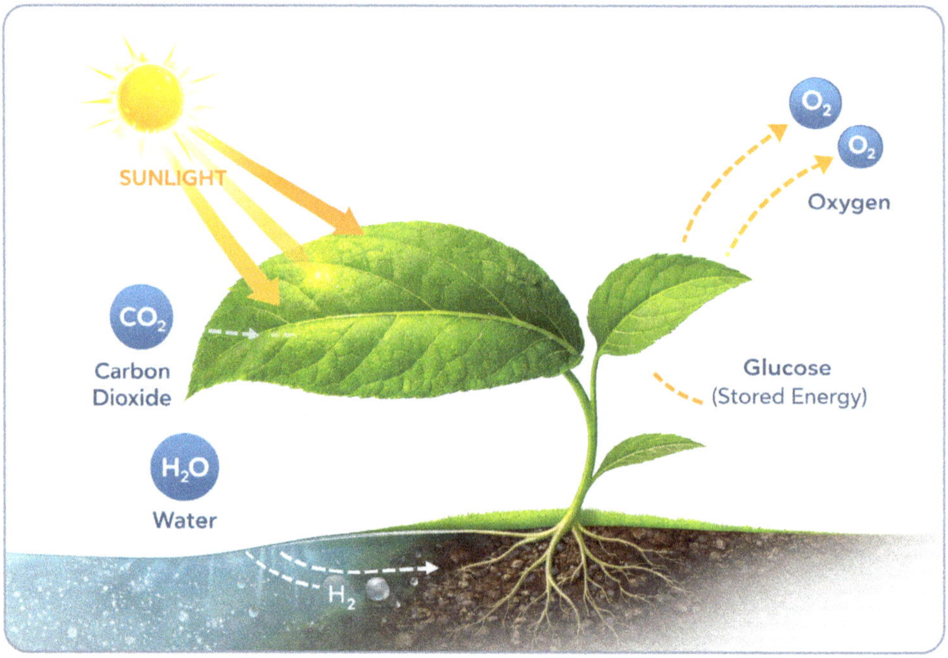

The transparency of water to visible light is equally significant. Because light can penetrate the upper layers of the ocean, photosynthetic organisms can thrive there. Plants and other photosynthetic organisms, such as ocean bacteria and algae, produce oxygen and form the foundation of marine ecosystems.

If water strongly absorbed visible light, life in Earth's oceans would be dramatically limited.

In reality, the physical properties of matter, the behavior of light, and the structure of Earth's environment work together in remarkable harmony.

LIGHT AND ITS NEAR-PERFECT EFFICIENCY IN PHOTOSYNTHESIS

Even more remarkable, recent research has revealed that photosynthesis operates with extraordinary efficiency at the quantum level, meaning at the scale of atoms and subatomic particles where energy behaves according to the laws of quantum physics.

When photons strike the light-harvesting structures inside plant cells, their energy must be transferred rapidly to specialized reaction centers where the energy can be converted into chemical energy.

> Life does not merely use light. It captures and directs it with remarkable precision.

Rather than move randomly through plant tissue, this energy appears to follow highly efficient pathways. Studies suggest that the transfer process may involve **quantum coherence** — a phenomenon in which energy simultaneously explores multiple possible routes and selects the most efficient path. This allows plants to capture and use incoming light with near-optimal efficiency. Quantum coherence also minimizes energy loss.

Such growth and energy use reflects a level of coordination far beyond simple chemistry. The molecular structures within photosynthetic systems appear precisely arranged to guide light's photon energy with remarkable effectiveness. Even at the smallest scales of biology, living systems demonstrate an extraordinary ability to harness light with precision.

The conversion of light into chemical energy during photosynthesis is therefore not merely functional but highly optimized. The processes that sustain life on Earth operate with efficiency approaching the limits permitted by physics.

Life does not merely use light — it captures and directs it with remarkable precision.

LIGHT AND INFORMATION AT THE CORE OF LIFE

DNA is not merely chemistry. **It is coded information written in a precise language of nucleotides.** Nucleotides are the molecular "letters" that form the genetic code, arranged in specific sequences that store biological instructions. The cell interprets this DNA information, copies it, repairs it, and transmits it across generations.

Life therefore consists of more than matter and energy. Life includes meaning.

Life is matter + energy + information.

From the earliest moment of development, biological systems exhibit coordinated processes directed toward function and organization. Such systems reflect not chaos, but structured order. These biological systems testify to God's brilliant creation.

Communication and coordination extend beyond individual human cells. Some researchers have shown that forests themselves form interconnected systems, where trees exchange chemical signals through underground fungal networks sometimes called the "wood wide web."

Such systems reveal that life is deeply structured and communicative at many levels. God is the author of this genius.

LIGHT AND HUMAN HEALTH

Light does more than initiate biological processes. Light continually sustains and regulates life. Living organisms depend on light not only for energy but for health, growth, repair, and biological timing.

Human physiology is closely synchronized with sunlight. We waken when it's light and sleep when it's dark. Circadian rhythms regulate sleep, metabolism, and behavior according to daily cycles of illumination.

Exposure to sunlight enables the body to produce vitamin D, a molecule essential for bone development, immune function, and cellular regulation. Without sufficient light exposure, numerous biological systems begin to fail.

Even at the molecular level, certain enzymes use light to repair damaged DNA, preserving the integrity of life's genetic code.

Modern medicine uses light as therapy. Carefully controlled wavelengths of light treat skin disorders, promote wound healing, reduce inflammation, and regulate seasonal mood disorders such as seasonal affective disorder (SAD).

Laser technology allows surgeons to perform extremely precise procedures, repair tissue, and restore vision.

Light is therefore not only a source of biological energy. Light is an instrument of healing.

Light exposure also influences immune function and cellular repair processes.

Certain wavelengths stimulate biological responses that help maintain the integrity of living tissues. The body's interaction with light reflects an ongoing relationship between physical energy and biological order.

Other living organisms —besides humans— are also synchronized with light. Many organisms use light signals to guide growth, navigation, and reproduction.

Circadian rhythms help regulate the body's daily cycles of sleep, wakefulness, and activity, guided by patterns of light and darkness.

LIGHT AND RECENT DISCOVERIES IN BIOELECTRIC COMMUNICATION

Recent research has revealed that human skin and other tissues communicate through electrical signals in ways previously unknown.

Scientists have discovered that injured skin cells generate slow electrical pulses that spread to neighboring cells, signaling damage and recruiting surrounding tissue for repair. These signals resemble neural communication but occur at a slower rate, demonstrating that electrical coordination is not limited to the nervous system but operates throughout the body.

Even more striking, researchers have found that tissues use bioelectric signals to coordinate collective cellular behavior. Electrical currents flowing across cell membranes help regulate growth, maintain tissue organization, and even identify damaged cells for removal.

Rather than act independently, cells appear to participate in coordinated electrical networks that maintain biological order.

Light plays an important role in many of these processes. Certain wavelengths of light influence cellular electrical activity, affect signaling molecules, and help regulate biological rhythms and repair mechanisms within tissues.

This emerging understanding of human skin and other tissues communicating through electrical signals reinforces a broader principle seen throughout biology: life is sustained not only by chemistry but by organized flows of energy, information, and electrical signaling.

From cellular energy to biological timing, light governs life at every scale.

A REFLECTION OF CREATION

This profound integration of energy, information, and order resonates with Scripture:

> *"For You formed my inward parts; You covered me in my mother's womb." – **Psalm 139:13***

Just as God spoke light into existence, new human life begins with directed information and energy, not chaos. Biology, at its core, is profoundly electrical and informational, a living echo of creation itself.

NOTICE AND DISCUSS

1. *Living organisms rely on light for many essential processes. Why is light so important for sustaining life on Earth?*
2. *Scientists have discovered that living cells use electrical and light-based signals to coordinate activity inside the body. What does this reveal about the complexity of life?*
3. *Photosynthesis captures sunlight with remarkable efficiency. Why might this process inspire both scientists and engineers?*
4. *Researchers describe the beginning of life as involving a "spark" of electrical or energetic activity within living cells. What might this "spark of life" suggest about when life truly begins?*
5. *Light plays a role in processes ranging from plant growth to cellular communication. What do these connections between light and life reveal about the design of living systems?*

“

You are the light of the world... Let your light so shine before men.

Matthew 5:14, 16

CHAPTER 9

Light in Scripture

FROM GENESIS TO CHRIST TO THE FUTURE

The first specific thing God speaks into existence in the creation account is:

> *"Let there be light."* – **Genesis 1:3**

This declaration is profound for several reasons:

First, light appears before the sun, moon, and stars. This means light is not dependent on heavenly bodies, but rather, those bodies later become instruments that bear and distribute the light God first created. Light itself originates from God.

Second, God separates light from darkness, naming them Day and Night. From the very beginning, light is associated with order, goodness, and structure. Darkness represents what is unformed and unstructured, but light brings definition, visibility, and function. Creation itself unfolds through this pattern: God speaks, light appears, and chaos gives way to order.

Scripture consistently ties light to God's presence, power, and character. Light reveals who God is and how He relates to His creation.

"God is light and in Him there is no darkness at all." – ***1 John 1:5***

Light is not merely something God produces; light reflects God's very nature. God's purity, holiness, and truth are described in terms of light. He dispels darkness.

Darkness symbolizes sin, deception, and disorder.

The Bible also connects light directly with life itself:

"In Him was life, and the life was the light of men." – ***John 1:4***

"The light shines in the darkness, and the darkness did not comprehend it." – ***John 1:5***

Here light represents more than physical illumination. Light signifies:

- spiritual understanding,
- moral clarity, and
- the very source of life.

> From Genesis to Revelation, light bookends God's story: the story of creation and redemption.

Darkness cannot overcome or extinguish this light. God has the ultimate triumph of truth over confusion and rebellion.

A simple illustration makes this clear. When a room is filled with light, you cannot shine darkness into it, instead you must remove the light to produce darkness. But when a room is dark and a light is turned on, the darkness cannot resist or oppose it, it vanishes instantly. In the same way, truth dispels error, and God's light overcomes darkness wherever it shines.

Throughout the Old Testament, divine encounters were often accompanied by radiant brightness.

When Moses met God on Mount Sinai, the mountain was covered with fire and glory (Exodus 24:16–17).

After speaking with God, Moses' face shone so brightly that it had to be veiled (Exodus 34:29–35).

God led Israel through the wilderness by a pillar of fire at night, providing both guidance and protection (Exodus 13:21). Light was a visible sign of divine presence.

In the Tabernacle and later the Temple, God's presence was associated with radiant glory, often called the *Shekinah*. The golden lampstand (menorah) continually burned, symbolizing God's ongoing presence among His people (Exodus 25:31–40).

The psalmist reflected this connection between God and illumination:

"The Lord is my light and my salvation; whom shall I fear?" – ***Psalm 27:1***

Light is therefore linked not only with revelation but with:

- guidance,
- safety, and
- deliverance.

Furthermore, the Bible also uses light as a picture of moral direction and truth. To walk in His ways is to walk in the light, to see reality correctly and live accordingly.

"Your word is a lamp to my feet and a light to my path." – ***Psalm 119:105***

"But the path of the just is like the shining sun, That shines ever brighter unto the perfect day.." – ***Proverbs 4:18***

Just a glance at the latest news serves as a constant reminder of the ongoing battle of light vs. darkness, good vs. evil that has persisted throughout human history. Without God's light to guide us, a worldview emerges in which moral boundaries blur, and people begin to call evil good and good evil.

"Woe to those who call evil good, and good evil; Who put darkness for light, and light for darkness; Who put bitter for sweet, and sweet for bitter!" – ***Isaiah 5:20***

The prophets anticipated a coming age when God's light would reach all nations. Isaiah foretold a future deliverer who would bring illumination to a darkened world:

"The people who walked in darkness have seen a great light." – ***Isaiah 9:2***

"I will also give You as a light to the Gentiles,That You should be My salvation to the ends of the earth– ***Isaiah 49:6***

These promises point forward to Jesus Christ, in Whom the symbolism of light reaches its fullest expression. Christ is not merely a teacher about light; He is Light in a human (incarnate) form. Jesus said:

"I am the light of the world. Whoever follows Me will not walk in darkness, but will have the light of life." – ***John 8:12***

Jesus does not simply reveal truth, He embodies it. In Him, divine light enters human history in visible form. His life, teachings, miracles, death, and resurrection illuminate both the nature of God and the condition of humanity.

Jesus being light continues at The Transfiguration, where Christ's face shone like the sun and His clothes become dazzling white (Matthew 17:2). This revealed to Peter, James and John that Jesus is God.

Jesus called His followers to reflect God's light:

"You are the light of the world... let your light shine before others." ***– Matthew 5:14–16***

Believers do not generate God's light themselves; they reflect the light of Christ, just as the moon reflects the sun.

From Genesis to Revelation, light bookends the story of creation and redemption. The Bible begins with the creation of light and ends with a restored creation filled entirely with God's light:

> *"The city had no need of the sun or of the moon to shine in it, for the glory of God illuminated it. The Lamb is its light."* – ***Revelation 21:23***

The final vision of Scripture returns to the same truth declared at the beginning:

- All light ultimately comes from God.
- Creation starts with light.
- Creation operates with light.
- Redemption is revealed through the Light of Jesus, and
- Eternity is illuminated by the Light of God Himself.

Light, therefore, is not merely a poetic symbol in Scripture. Light is a created reality that reflects divine order, reveals truth, and sustains life. From God's first command of creation to the eternal kingdom, light testifies to God's design, His presence, and His plan to bring light, clarity, and life to a world once shrouded in darkness.

NOTICE AND DISCUSS

1. *Where does the Bible first introduce light, and why might light appear before the Sun and stars in the Genesis account?*
2. *In what ways does Scripture use light as a symbol for truth, life, and God's presence?*
3. *Jesus said, "I am the Light of the world." What do you think this statement means?*
4. *Why do you think light appears so frequently throughout the Bible, from Genesis to Revelation?*
5. *How does understanding the science of light deepen your appreciation for the biblical imagery of light?*

“

Your word
is a lamp
to my feet
And a light
to my path.

PSALM 119:105

CHAPTER 10

Jim Lovell and the Light in the Darkness

During the Korean War, future Apollo 13 astronaut Jim Lovell faced a life-threatening crisis.

Flying at night over the ocean, all his cockpit lights suddenly failed. With no visible horizon and dwindling fuel, he could not locate the aircraft carrier he needed to land on. Surrounded by darkness and unable to see the ship, he faced the very real possibility of running out of fuel and crashing into the sea.

Then he noticed something extraordinary.

Light does more than make objects visible, it makes knowledge possible.

The ocean behind the aircraft carrier glowed faintly with **bioluminescence,** living light produced by microscopic marine organisms disturbed as the vessel moved through the water. As the carrier cut through the dark ocean, it left a glowing trail behind it. That hidden, living light revealed the ship's position and guided Lovell safely home.

In one of Lovell's darkest moments, light embedded within God's creation preserved his life.

Bioluminescence itself is a remarkable phenomenon. Many living organisms, from tiny plankton to deep-sea creatures to worms on land, produce light through precisely regulated chemical reactions. Even in the depths of the ocean, far from sunlight, life generates its own illumination. Creation is filled not only with reflected light, but with living light.

Lovell's experience reveals an important truth about the nature of light. Without light, the world becomes disorienting and unknowable. In darkness, direction is lost, distance cannot be judged, and reality itself becomes hidden. Light does more than make objects visible, it makes knowledge possible.

Lovell's experience therefore illustrates more than a dramatic rescue. It reflects a deeper pattern woven throughout creation: light reveals, guides, and makes understanding possible.

Scripture expresses this same reality:

"And the light shines in the darkness, and the darkness did not comprehend it." – ***John 1:5***

Light often appears where we least expect it, even in the darkest circumstances. From the stars in the heavens to the cells of living organisms, and even to the hidden glow of bioluminescence in the ocean at night, light reveals, guides, and sustains. When darkness surrounds us, light is never absent. It is often simply waiting to be seen.

“

For you were once darkness, but now you are light in the Lord. Walk as children of light.

EPHESIANS 5:8

CHAPTER 11

Walk as Children of Light

The Bible does not call us merely to admire light — it calls us to embody it.

> *"You were once darkness, but now you are light in the Lord. Walk as children of light."* – ***Ephesians 5:8***

Throughout this book we have seen how the life-giving power of light permeates reality.

Light shapes the cosmos, sustains life, carries information, and reveals hidden order within creation.

Yet the Bible teaches that light is not only a physical reality, it is also a moral and spiritual calling.

To walk in light is to walk in truth, humility, and love.

Walking in light means rejecting deception, seeking wisdom, and reflecting Christ's character in daily life.

Just as physical light reveals what is hidden, spiritual light exposes falsehood and brings clarity to the human heart.

Light illuminates.

Light reveals.

Light gives life.

So too, those who believe the gospel of Christ are called to reflect these qualities in the world.

WALK IN TRUTH

Light represents truth. Darkness conceals, but light reveals reality as it truly is.

To live as children of light is to pursue honesty, integrity, and understanding. It means loving what is true even when living by that truth is difficult.

Light means allowing God's light to examine our own lives.

WALK IN LOVE

Light also gives life. Just as sunlight sustains the living world, love sustains human relationships.

To walk in light is to act with compassion, mercy, kindness, goodness, temperance, and selflessness toward others.

Walking in light reflects the character of Jesus who called Himself "the light of the world."

WALK IN WISDOM

Light provides guidance.

A path illuminated by light allows safe passage, while darkness leads to confusion and danger.

Just as daily cycles of light help regulate the rhythms of life, God's truth helps guide the rhythms of a life lived in wisdom.

The Bible repeatedly describes God's Word as a lamp that directs our steps. Walking in light therefore means seeking wisdom and living according to it. This is God's design.

REFLECT THE LIGHT

Light does not exist merely for itself — it radiates outward. In the same way, believers are called to reflect God's loving light to others.

> *"Let your light shine before men, that they may see your good works and glorify your Father in heaven."* – **Matthew 5:16**

And while salvation is attained by simple belief, living the Christian life to the fullest is more than just belief, it is a life transformed from darkness into light

Those who receive light are called to become sources of light.

A RADICAL WAY OF LIVING

In a world filled with confusion, division, and moral uncertainty, walking in light is a radical act. It always has been.

Walking in light means choosing clarity over chaos, truth over deception, and love over selfishness.

> In case you've never heard the gospel that saves souls: There is a God, and you are not Him. He is the righteous God of the Bible who created the world, and one day you will meet Him. The reason we die is because we are sinners. But in His mercy, God sent His Son, Jesus Christ, to die for your sins and mine. He rose from the dead to freely offer us forgiveness and eternal life. If you will trust in Him and believe that He can forgive your sins, He will. Do you believe?

Walking in light means living in a way that reflects the order and meaning woven into creation itself.

Light is not only something we observe — it is something we are called to become.

From the first words of creation — "Let there be light" — to the promise of the New Creation where God Himself will be our light, the Bible reveals a universe illuminated by truth, order, and purpose.

Light fills the cosmos, sustains life, and reveals the beauty of creation. Yet the greatest light is not found in the stars or the Sun. It is found in Jesus Christ, the Light of the world. As we walk in His light, we reflect that light to others, pointing them to the Creator who spoke light into existence in the beginning.

NOTICE AND DISCUSS

1. *The Bible often contrasts light and darkness. What do these images represent in Scripture?*
2. *Jesus said, "You are the light of the world" (Matthew 5:14). What do you think it means to live as a person of light?*
3. *How can truth, honesty, and integrity reflect the light that Scripture describes?*
4. *In what ways can believers bring light into a world that often feels dark or confused?*
5. *What lesson from this book about light – scientific or biblical – will stay with you the longest?*

“

But the
Lord will be
to you an
everlasting
light…

Isaiah 60:19

CONCLUSION

Eternal Light

THE NEW CREATION

The Bible ends with a breathtaking vision:

"There shall be no night there... for the Lord God gives them light."
*– **Revelation 22:5***

In the New Creation, light will no longer come from the sun or stars, it will flow directly from God Himself. What began with the command "*Let there be light*" will culminate in eternal light.

No darkness.

No shadows.

No artificial illumination.

Only the radiant presence of God.

Light stands at both the beginning and the end of creation's story.

Yet light reveals more than physical reality; light reveals purpose.

From the first light photon in creation to the spark of life in the womb, from quantum mysteries to the cosmic background radiation, light continually points toward intelligence, order, and meaning. We recognize this as our intelligent God who creates order and meaning that matches us perfectly.

Throughout this book we have seen that light is a window into the nature of reality itself.

Jesus is not merely a teacher about light; He is Light incarnate. To trust Jesus is to walk in truth, clarity, and life. To study light is therefore more than a scientific pursuit; it is an invitation to glimpse the mind of the Creator.

But the Bible also tells us why we need this light. Humanity has turned away from God, and sin has brought darkness into the human heart. Yet God, in His mercy, sent His Son into the world so that those who trust in Him would not remain in darkness. Through Jesus' death and resurrection, forgiveness and new life are offered freely to all who accept His gift of salvation, a salvation that is not earned by our works but received by grace through faith.

As the Bible declares:

> *"For it is the God who commanded light to shine out of darkness, who has shone in our hearts to give the light of the knowledge of the glory of God in the face of Jesus Christ."* – **2 Corinthians 4:6**

WHAT LIGHT REVEALS

Throughout this book we have seen that light is not merely a physical phenomenon, but a window into the nature of reality itself:

- Light dispels darkness
- Light cannot be grasped or held
- Light can be everywhere at once
- Light is not consumed by what it illuminates
- Light enables us to see and find our way
- Light communicates information to the mind, including science
- Light reveals color and beauty
- Light provides warmth and comfort
- Light produces energy and power
- Light travels with extraordinary speed
- Light enables growth and healing
- Light fades the farther we move from its source

Throughout this book we have seen that light is woven into the very fabric of creation. It reveals structure in the cosmos, powers life on Earth, and carries information across the universe. Light is not accidental. It is part of a creation filled with order, meaning, and purpose.

The Bible tells us that one day the need for created lights will pass away, because God Himself will illuminate the new creation. The Lamb will be its light.

The story of light is ultimately the story of creation, redemption, and restoration — a universe illuminated by the God who is Himself the eternal Light.

Until that day, every beam of light we see continues to remind us of the One who speaks light into existence.

APPENDIX A

Key Scripture Passages

CREATION AND THE NATURE OF LIGHT

- **Genesis 1:3-5** – Creation of light and dividing light from darkness
- **Isaiah 45:7** – God forms light and creates darkness
- **Psalm 104:2** – God clothes Himself with light
- **Daniel 2:22** – Light dwells with God
- **2 Corinthians 4:6** – God commanded light to shine out of darkness
- **1 Timothy 6:16** – God dwells in unapproachable light

CHRIST AS THE LIGHT

- **John 1:1-9** – Christ as the true Light
- **John 3:19-21** – Light has come into the world
- **John 8:12** – Jesus is "the light of the world"
- **John 9:5** – Jesus is the Light of the world while in the world
- **John 12:46** – Christ came as light into the world so we people would not choose to live darkness
- **Luke 2:32** – Light for revelation to the Gentiles (anyone who is not Jewish)
- **Hebrews 1:3** – Jesus is the radiance of God's glory

GOD'S PRESENCE, GLORY, AND REVELATION

- **Exodus 13:21–22 –** God led the people in a pillar of fire
- **Exodus 24:16–17 –** The glory of the Lord like consuming fire
- **Exodus 34:29–35 –** Moses' face shines after meeting God
- **Psalm 27:1 –** The Lord is my light and salvation. Whom shall I fear?
- **Psalm 36:9 –** In Your light we see light
- **Habakkuk 3:4 –** God's radiance like the sunlight

LIGHT AS TRUTH, WISDOM, AND GUIDANCE

- **Psalm 19:1 –** The heavens declare God's glory
- **Psalm 119:105 –** God's Bible is a lamp to my feet
- **Psalm 119:130 –**God's words give light and understanding
- **Proverbs 6:23 –** God' commandments are a lamp
- **Proverbs 4:18 –** The path of the righteous shines brighter
- **2 Peter 1:19 –** God's word is a lamp in a dark place during all our lives

BELIEVERS AS CHILDREN OF LIGHT

- **Matthew 5:14-16 –** God-followers are the light of the world
- **Ephesians 5:8 –** Walk as children of light
- **Philippians 2:15 –** Shine blameless and harmless as lights in the world
- **1 Thessalonians 5:5 –** God-followers are sons of light who do not follow night or darkness
- **1 John 1:5–7 –** Walk in the light

ETERNAL AND FUTURE LIGHT

- **Isaiah 9:2** – A great light for those in darkness or who had been in darkness
- **Isaiah 60:19-20** – The Lord as everlasting light and end of mourning
- **Revelation 21:23-25** – God's glory gives light to the city in heaven
- **Revelation 22:5** – God as eternal light

APPENDIX B

Scientific & Historical References

LIGHT & OPTICS

- Feynman, Richard P. QED: *The Strange Theory of Light and Matter.* Princeton University Press, 1985.
- Hecht, Eugene. Optics. Pearson Education.
- Maxwell, James Clerk. A *Treatise on Electricity and Magnetism.* Oxford University Press, 1873.
- Newton, Isaac. *Opticks.* 1704.
- Young, Thomas. "On the Nature of Light and Colours." *Philosophical Transactions of the Royal Society,* 1802.

QUANTUM PHYSICS & INFORMATION

- Einstein, Albert. "On a Heuristic Point of View Concerning the Production and Transformation of Light." *Annalen der Physik,* 1905.
- Feynman, Richard. *The Character of Physical Law.* MIT Press, 1965. Bell, John S. "On the Einstein-Podolsky-Rosen Paradox." *Physics,* 1964.
- Aspect, Alain, Dalibard, J., & Roger, G. "Experimental Test of Bell's Inequalities." *Physical Review Letters,* 1982.

- Schrödinger, Erwin. *Collected Papers on Wave Mechanics.*
- Wheeler, John Archibald. "Information, Physics, Quantum: The Search for Links."
- Lloyd, Seth. *Programming the Universe: A Quantum Computer Scientist Takes on the Cosmos.* Knopf, 2006.
- Lloyd, S. "Computational Capacity of the Universe." *Physical Review Letters*, 2002.
- Tegmark, Max. "The Mathematical Universe." *Foundations of Physics*, 2008.
- Davies, Paul. *The Demon in the Machine: How Hidden Webs of Information Are Solving the Mystery of Life.* University of Chicago Press, 2019.
- Meyer, Stephen C. Signature in the Cell. HarperOne, 2009.
- Shannon, Claude. "A Mathematical Theory of Communication." *Bell System Technical Journal*, 1948.
- Wiener, Norbert. *Cybernetics: Or Control and Communication in the Animal and the Machine.* MIT Press.
- Scientific American Guest Blog. "What Does the New Double-Slit Experiment Actually Show?" https://blogs.scientificamerican.com/guest-blog/what-does-the-new-double-slit-experiment-actually-show/

COSMOLOGY & FINE-TUNING

- NASA Cosmic Background Explorer (COBE) Mission Results, 1990–1993.
- Wilkinson Microwave Anisotropy Probe (WMAP) Mission Results, NASA.

- Planck Satellite Mission Results, European Space Agency.
- Peebles, P. J. E. *Principles of Physical Cosmology*. Princeton University Press.
- Sloan Digital Sky Survey (SDSS) Data Releases and Large-Scale Structure Observations.
- Gonzalez, Guillermo, & Richards, Jay. *The Privileged Planet*. Regnery Publishing.
- Ward, Peter D., and Brownlee, Donald. Rare Earth: Why Complex Life Is Uncommon in the Universe. Springer, 2000.

QUANTIZED REDSHIFT & ALTERNATIVE COSMOLOGY

- Setterfield, Barry. Research on zero-point energy and variable speed of light models. https://barrysetterfield.org/
- Spears, Joe. "Plasma Astronomy: A Different View." TASC Creation Science. https://tasc-creationscience.org/article/plasma-astronomy-different-view
- Spears, Joe. "Cosmic Electric Currents." TASC Creation Science. Discussion of reported large-scale electric currents associated with astrophysical jets and related cosmic structures, with cited supporting sources. https://tasc-creationscience.org/article/plasma-astronomy-different-view-part-2
- Spears, Joe. "Quantized Redshift Explained." TASC Creation Science. https://tasc-creationscience.org/article/quantized-redshift-explained
- Humphreys, D. Russell. "Quantized Redshifts in the Universe." Creation Research Society publications and related papers.

- Simhony, Yaakov. "The Electron-Positron Lattice Model of Space (EPOLA)." Various publications and papers on structured vacuum models.
- Tifft, William G. "Discrete States of Redshift and Galaxy Dynamics." *Astrophysical Journal.*
- Arp, Halton. *Seeing Red: Redshifts, Cosmology and Academic Science.* Apeiron.
- Anderson, Carl D. "The Positive Electron." *Physical Review* 43 (1933): 491–494.

PLASMA PHYSICS & ELECTRIC UNIVERSE PERSPECTIVES

- Alfvén, Hannes. *Cosmic Plasma.* Reidel Publishing.
- Birkeland, Kristian. *The Norwegian Aurora Polaris Expedition* 1902–1903.
- Langmuir, Irving. "Oscillations in Ionized Gases." *Proceedings of the National Academy of Sciences.*
- Peratt, Anthony. *Physics of the Plasma Universe.* Springer.
- Scott, Donald E. *The Interconnected Cosmos.* Mikamar Publishing.
- Juergens, Ralph. "The Sun as an Electric Discharge." *Kronos Journal.*

SOLAR PHYSICS & SPACE SCIENCE

- NASA Solar Dynamics Observatory (SDO) Mission Data.
- Parker Solar Probe Mission Findings, NASA.
- McHenry, Ellen. *Sun Puzzles and Solar Mysteries.*
- Phillips, Kenneth J. H. *Guide to the Sun.* Cambridge University Press.

BIOLOGY, BIOELECTRICITY & BIOPHOTONS

- Haddock, S.H.D., Moline, M.A., & Case, J.F. "Bioluminescence in the Sea." *Annual Review of Marine Science*, 2010.
- Kobayashi, M. et al. "In Vivo Imaging of Human Ultraweak Photon Emission." PLOS ONE, 2009.
- Popp, Fritz-Albert. *Biophotonics: The Study of Light Emission from Biological Systems.*
- Levin, Michael. "Bioelectric Signaling in Regeneration and Development." *Annual Review of Cell and Developmental Biology.*
- Jaffe, Lionel F. "The Role of Bioelectric Currents in Development." *Cold Spring Harbor Symposia on Quantitative Biology.*
- Pollack, Gerald H. The Fourth Phase of Water: Beyond Solid, Liquid, and Vapor. Ebner & Sons, 2013.
- Pollack, Gerald H. "Water, Energy and Life." International Journal of Design & Nature, 2010.
- Collini, Elisabetta, et al. "Coherently Wired Light-Harvesting in Photosynthetic Marine Algae at Ambient Temperature." Nature, 463 (2010): 644–647.
- Saey, Tina Hesman. "Skin Cells Emit Slow Electrical Pulses After Injury." Science News, April 2, 2025.
- Cutts, Elise. "Cells Use 'Bioelectricity' to Coordinate and Make Group Decisions." Quanta Magazine, January 12, 2026.

MEASUREMENT, TIME, AND PHYSICAL CONSTANTS

- Essen, Louis. *The Measurement of Time*. Oxford University Press.
- International Bureau of Weights and Measures (BIPM). *The International System of Units (SI)*.
- National Institute of Standards and Technology (NIST). Definition of the Meter and Speed of Light.
- Sheldrake, Rupert. *The Science Delusion: Freeing the Spirit of Enquiry*. Coronet, 2012.

APPENDIX C

Glossary

- **Atomic Clock** – A highly precise clock that measures time using the natural vibration frequencies of atoms.
- **Bioluminescence** – The production of light by living organisms through chemical reactions within their bodies. Living light produced by microscopic marine organisms disturbed as vessels move through water (See page 83).
- **Biophotons** – Extremely weak light emitted by living organisms as a result of metabolic processes within cells. Examples of these processes include oxidative reactions, cellular respiration, and other biochemical reactions that release small amounts of light energy.
- **Circadian Rhythm** – Biological processes, such as sleep, waking, eating, working, that follow a roughly 24-hour cycle regulated by light and darkness.
- **Cosmic Microwave Background (CMB)** – Faint radiation observed throughout the universe, commonly understood as a remnant of the early universe and studied to understand cosmic structure.
- **Design** – The appearance of purposeful arrangement or intentional structure in nature, often interpreted as evidence of an intelligent cause.
- **Electromagnetic Spectrum** – The full range of light wavelengths, from radio waves to gamma rays, including the small portion visible to the human eye. See the visible spectrum in the illustration on page 14.

- **Electron** – A negatively charged subatomic particle found surrounding the nucleus of an atom. Electrons play a key role in electricity, chemistry, and the behavior of light and plasma.
- **Energy** – The capacity to do work or produce change. Light carries energy that can produce heat, power chemical reactions, transmit signals, and drive many of the processes that sustain life.
- **Fine-Tuning** – The observation that the physical laws and constants of the universe appear precisely suited to allow life and complex structures.
- **Frequency** – The number of wave cycles that pass a point in a given time. In light waves, frequency determines the energy of the radiation.
- **Ion** – An atom or molecule that has gained or lost one or more electrons, giving it an electrical charge. Positively charged ions have lost electrons, while negatively charged ions have gained electrons. Ions are common in plasmas, lightning, and many chemical reactions.
- **Information (Physics)** – Structured, meaningful content carried by signals or physical systems. Light transmits information through its frequency, intensity, and pattern.
- **Intellectual phase locking** – A metaphor comparing scientific consensus to physical phase locking, where systems synchronize to a dominant signal. In this context it refers to the tendency of research communities to align around prevailing theories, sometimes resisting alternative interpretations.
- **Intellectual phase-prestidigitation** – Kind of scientific sleight of hand that locks in the answer before the experiment even begins.
- **Neutrino** – An extremely light subatomic particle that rarely interacts with matter. Vast numbers of neutrinos pass through Earth and even through our bodies every second.

- **Neutron** – A neutral subatomic particle found in the nucleus of an atom. Neutrons contribute to atomic mass but carry no electric charge.
- **Nucleotides** – The basic molecular units that make up DNA and RNA. They function like letters in the genetic code that store biological instructions.
- **Photon** – A discrete packet of light energy. Light is detected and absorbed in individual units called photons.
- **Photosynthesis** – The process by which plants capture light energy and convert it into chemical energy used to sustain life.
- **Plasma** – An electrically charged state of matter consisting of ions and free electrons, often called the fourth state of matter. As far as we can discern, stars and much of the universe are composed of plasma.
- **Proton** – A positively charged subatomic particle located in the nucleus of an atom. The number of protons determines the identity of an element.
- **Quantized Redshift** – The idea that changes in the wavelength of light from distant objects may occur in discrete steps rather than continuously.
- **Quantum Coherence** – A quantum phenomenon in which energy exists in multiple possible states at once and remains coordinated, allowing it to follow several possible paths and move through a system very efficiently.
- **Quantum Entanglement** – A quantum phenomenon in which two particles remain linked so that measuring one instantly determines the state of the other, even when separated by great distances.
- **Quantum Tunneling** – A quantum process in which particles pass through barriers that classical physics predicts they cannot cross.

- **Quark –** A fundamental particle that combines with other quarks to form protons and neutrons, the particles found in atomic nuclei.
- **Redshift –** The stretching of light toward longer wavelengths, commonly observed in light from distant astronomical objects.
- **Spectral Lines –** Distinctive patterns of light wavelengths emitted or absorbed by elements, used to identify chemical composition and motion of stars and galaxies.
- **Spectrum (Light Spectrum) –** The range of colors or wavelengths produced when light is separated into its components.
- **Triality (of Light) –** The concept that light exhibits three complementary aspects: wave behavior, particle behavior, and information-carrying structure.
- **Wave-Particle Duality –** The discovery that light behaves both as a wave and as a particle depending on how it is observed or measured.
- **Wavelength –** The distance between repeating peaks of a wave, determining the color or energy of light.
- **Zero-Point Energy –** The background energy present even in empty space due to constant quantum fluctuations.

APPENDIX D

Read More!

SCRIPTURE & BIBLICAL STUDY

- **The Holy Bible** — especially Genesis 1, Ephesians, Psalm 19, John 1, John 8, and Revelation 21–22.
- BibleGateway.com — searchable Scripture in multiple translations.
- BlueLetterBible.org — original language study tools.

REAL SCIENCE RADIO RESOURCES

- Real Science Radio (rsr.org, YouTube.com/@RealScienceRadio) — science news, interviews, and research from a biblical perspective.
- store.rsr.org — books, DVDs, and educational materials.

CREATION SCIENCE RESOURCES

- Creation Science Center — creationscience.com
- Creation Ministries International — Creation.com
- Institute for Creation Research — ICR.org
- Creation Research Society — CreationResearch.org
- Answers in Genesis — AnswersInGenesis.org

BOOKS FOR FURTHER READING

- Feynman, Richard. QED: *The Strange Theory of Light and Matter.*
- Sheldrake, Rupert. *The Science Delusion.*
- McHenry, Ellen. *Sun Puzzles.*
- McHenry, Ellen. *Plasma Astronomy and the Bible.*

ABOUT REAL SCIENCE RADIO

Real Science Radio (RSR) is a Christian apologetics ministry dedicated to presenting evidence for God's design in the natural world.

Real Science Radio examines scientific discoveries in light of Scripture and challenges materialistic explanations of origins, life, and the universe.

Real Science Radio equips believers to think critically about scientific claims and defend the biblical worldview.

Real Science Radio is co-hosted by **Fred Williams** and **Doug McBurney**, who demonstrate the harmony of science and biblical faith.

Through engaging discussions and research, the program addresses topics such as:

- physics
- biology
- cosmology
- geology
- philosophy

including how various areas of science present evidence of the design, purpose, and reliability of God's Word and world.

Real Science Radio broadcasts weekly in the Denver area on **50,000-watt AM 670 KLTT**, reaching listeners across the Rocky Mountain region and into nearby states. The program currently airs Fridays at 12:30 PM MST (see rsr.org for updates). RSR also reaches a global audience through its YouTube channel, along with its website, educational resources, and public presentations.

Real Science Radio is committed to:

- proclaiming the Gospel of Jesus Christ
- revealing evidence of design in creation
- defending the authority of Scripture
- equipping believers to engage culture with truth

Creation itself testifies to its Creator, and Real Science Radio seeks to make that testimony clear. Through decades of conversations with scientists and researchers, RSR presents perspectives that challenge conventional assumptions and encourage fresh thinking about the natural world.

By examining evidence through a biblical worldview, Real Science Radio points to wisdom grounded in Scripture rather than the changing ideas of secular culture.

> *"For the wisdom of this world is foolishness with God. "* — ***1 Corinthians 3:19***

- Learn more: **rsr.org**
- Watch and subscribe: **YouTube.com/@RealScienceRadio**
- Additional resources: **store.rsr.org**

ABOUT THE REAL SCIENCE RADIO TEACHING SERIES

The **Real Science Radio Teaching Series** presents scientific topics through a biblical worldview. Each book explains important discoveries in science and shows how the evidence points to order, meaning, and intentional design in creation.

These books are designed to be accessible to a wide audience and are especially useful for:

- homeschool science
- high school apologetics
- church discussion groups
- curious readers interested in science and faith

Through clear explanations, illustrations, and discussion questions, the series encourages readers to think carefully about the evidence and consider what the natural world reveals about its Creator.

We equip believers to think critically about scientific claims and show that the Bible is a true, trustworthy standard for science and life.

Additional titles in the Real Science Radio Teaching Series are in development.

- Learn more: **rsr.org**
- Watch and subscribe: **YouTube.com/@RealScienceRadio**
- Additional resources: **store.rsr.org**

Stars shine through towering clouds of gas in the Eagle Nebula, known as the "Pillars of Creation." Across the universe, light reveals the structure and beauty of the cosmos. Image credit: NASA / ESA / Hubble Space Telescope

www.ingramcontent.com/pod-product-compliance
Lightning Source LLC
LaVergne TN
LVHW070854210926
848482LV00016B/932